Alternating Currents

Electricity Markets and Public Policy

Timothy J. Brennan
Karen L. Palmer
and Salvador A. Martinez

Resources for the Future | Washington, DC

Printed in the United States of America

An RFF Press book
Published by Resources for the Future
1616 P Street, NW, Washington, DC 20036–1400
www.rff.org

Library of Congress Cataloging-in-Publication Data
Brennan, Timothy J.
Alternating currents : electricity markets and public policy / Timothy J. Brennan, Karen L. Palmer, and Salvador A. Martinez
p. c.m.
Includes bibliographical references and index.
ISBN 1-891853-52-X (lib. bdg.) — ISBN 1-891853-07-4 (pbk.)
1. Electric utilities—United States. 2. Electric utilities—Government policy—United States. 3. Electric utilities—United States—Management. I. Palmer, Karen L. II. Martinez, Salvador A. III. Title.

HD9685.U5 B74 2002
333.793′2′0973—dc21 2002017328

f e d c b a

The paper in this book meets the guidelines for permanence and durability of the Committee on Production Guidelines for Book Longevity of the Council on Library Resources.

The text of this book was designed and typeset by Betsy Kulamer in Utopia and Ocean Sans. It was copyedited by Alfred F. Imhoff. The cover was designed by Meadows Design Office, Inc.

ISBN 1–891853–07–4 (paper) ISBN 1–891853–52–X (cloth)

About *Resources for the Future* and *RFF Press*

Resources for the Future (RFF) improves environmental and natural resource policymaking worldwide through independent social science research of the highest caliber.

Founded in 1952, RFF pioneered the application of economics as a tool to develop more effective policy about the use and conservation of natural resources. Its scholars continue to employ social science methods to analyze critical issues concerning pollution control, energy policy, land and water use, hazardous waste, climate change, biodiversity, and the environmental challenges of developing countries.

RFF Press supports the mission of RFF by publishing book-length works that present a broad range of approaches to the study of natural resources and the environment. Its authors and editors include RFF staff, researchers from the larger academic and policy communities, and journalists. Audiences for RFF publications include all of the participants in the policymaking process—scholars, the media, advocacy groups, nongovernmental organizations, professionals in business and government, and the general public.

Contents

Preface vii

About the Authors xi

Introduction

1. Issues in Restructuring the Electricity Industry 1

Part I: How the Industry Got Here

2. Understanding the Electricity Industry 13
3. From Regulation to Competition 26
4. International and U.S. Restructuring Experiences 33
5. The California Experience 46

Part II: Current Policy Issues

6. Competition in Energy, Regulation of Wires 61
7. Vertical Restructuring 71
8. Regulating Rates for Transmission and Distribution 81
9. Encouraging Competition 92
10. Balancing Loads and Dispatching Power 106
11. Ensuring Reliability in a Competitive Market 116
12. State and Federal Roles 127
13. Public Power's Role after Restructuring 136

14. Covering Stranded Costs 149

15. Restructuring and Environmental Protection 159

16. Public Purpose Programs in a Competitive Market 173

Part III: The Future

17 . Prospects for Restructuring 187

Supplemental Reading 198

Index 205

Preface

In the spring of 1999, we and our colleagues at Resources for the Future began to think that it was time to consider updating *A Shock to the System*, the primer on electricity restructuring that RFF published in the summer of 1996. That book, which Tim Brennan and Karen Palmer coauthored with Dallas Burtraw, Ray Kopp, Alan Krupnick, and Vito Stagliano, was designed to introduce the subject of electricity restructuring to the policy community, business analysts, citizens' groups, educators, journalists, and anyone interested in how this complicated sector might undergo the intricate transition from regulation to competition.

Three years after the publication of *A Shock to the System*, much had changed. Some terms of art, such as calling a central electricity market a "PoolCo," had disappeared from industry parlance. Regulators and policymakers at the state and federal levels were coping with unprecedented numbers of mergers between electricity suppliers, including traditional utilities and independent power producers. At the national level, we saw expanding debates about the role of the federal government in encouraging states to open retail electricity markets and ensuring sufficient separation between generation and transmission to promote competition. Numerous issues, including system reliability, the role of public power, and redesigning public benefits programs (such as subsidies for conservation, research, and low-income household energy use) were receiving far more attention than they received in *Shock*.

Most importantly, the United States and other countries began to accumulate some experience with a more competitive electricity sector. The United Kingdom and Chile were among the leaders internationally in opening electricity markets, in many ways ahead of the United States. Since *Shock* was published, a multitude of states, among them California and Pennsylvania, began their own experiments with electricity competition. Market developments that had been largely speculative in the mid-1990s became real by the end of the decade.

Our initial assessments for whether a new primer on electricity competition was worthwhile, and what such a volume should contain, began to take shape through two very useful meetings. The first, held at RFF on May 10, 1999, was attended by a number of industry experts who deal with business planning, public policy, and environmental management. The second meeting, on July 13, 1999, was of the Electricity Working Group in Washington, an informal gathering of economists, engineers, and policy analysts in government, academia, think tanks, and private industry. Encouraged by the ideas presented in these meetings, we decided that electricity restructuring had evolved to a sufficient degree that we had enough material for an entirely new volume. We spent the next year or so putting that new volume together and sent it out for independent review, hoping to publish some time in early 2001.

Then, along came the California electricity crisis.

This sent us, along with most other observers of electricity restructuring, back to our drawing boards. We added extensive discussions of the California experience and the issues it brought to the forefront (particularly market power and the design of institutions for trading electricity). However, the lessons to be learned from California's woes are limited. In many respects, the California situation was the unfortunate result of a variety of idiosyncratic policy decisions. These turned an otherwise short-term supply crunch into a long and complex financial disaster and political morass that, among other things, made the supply crunch even worse.

This new volume, *Alternating Currents: Electricity Markets and Public Policy*, summarizes our thoughts regarding the major concerns now facing those charged (so to speak) with deciding whether and how to expand competition in electricity markets. As with *Shock*, our foremost goal is to provide an independent, objective assessment of these issues. Many who want to learn about electricity restructuring come to the subject without prior knowledge of its technology, policy developments, and the economic tools most helpful for understanding how markets perform. We hope our presentation can fill those gaps, providing an accessible guide to policymakers—and the public on whose behalf they work—as they consider how to design electricity markets, promote competition, ensure reliability, and protect the environment.

A book such as this is inevitably aimed at a moving target. Even as we write, electricity restructuring continues to evolve. The electricity crisis in California has abated, and Enron, one of the leading electricity marketers, has imploded. Some states have reconsidered restructuring, while others are continuing to go forward. We hope that the methods for and principles of understanding electricity markets presented here will help to illuminate developments that we did not foresee and discuss.

One other difference between now and when *A Shock to the System* was published is that, in 1996, there was little information available to the public on restructuring. Particularly after California's electricity crisis dominated the headlines from the summer of 2000 through the spring of 2001, a great deal of material is available. We have included a list of books, articles, periodicals, and Internet websites that can provide information to those seeking more details on topics of their choice.

Acknowledgements

Alternating Currents would not be here without the financial and intellectual support of Resources for the Future. Special thanks go to Paul Portney, Ray Kopp, and Alan Krupnick for arranging the financial support and for tremendous help with the substance and focus of the book. Numerous discussions with Dallas Burtraw and Howard Gruenspecht added immensely to our knowledge of what was going on in the industry and how to see the forest instead of getting lost in the trees. Ranjit Bharvirkar, Meghan McGuinness, and Peter Nelson provided extensive research assistance. Winston Harrington supplied the title as the winning entry in a contest among the RFF staff. (We're mum on the other entries, except to say that some of us were particularly fond of using the Who's "Talkin' 'bout My Generation.") We also want to thank Don Reisman, Gina Armento, and Rebecca Henderson of RFF Press for their careful attention, solicitation of very helpful anonymous outside reviewers, and patience in holding us to deadlines we'd rather have ignored.

Among those not part of the immediate RFF family, very special thanks go to Doug Hale of the U.S. Department of Energy's Energy Information Administration. In addition to the extensive help Doug gave us directly, his efforts as organizer of the aforementioned Electricity Working Group were an immensely valuable service in developing and sharing research on electricity economics and policy. His praises deserve to be sung from higher mountains than this, but it's a start. Among the participants in the Electricity Working Group, Udi Hellman, David Hunger, James Mietus, Paul Sotkewicz, and Peter Van Doren deserve particular mention.

Portions of this book, particularly sections relating to the California crisis, benefited greatly from presentations before a variety of academic and industry audiences. Special thanks go to John Siegfried, Luke Froeb, and the Owen School of Business at Vanderbilt University, and to Joseph Doucet and the Center for Applied Business Research in Energy and the Environment at the School of Business of the University of Alberta, for invited presentations on lessons to be learned from the California situation. We are grateful to Michael Crew and the Rutgers University Center for Research in Regulated Industries for the opportunity to present research on California and "convergence mergers" between electricity generators and natural gas suppliers. Participants in those conferences and at seminars at the Department of Energy, Environmental, and Mineral Economics at Pennsylvania State University; George Mason University; the Mansfield Center for Pacific Affairs; and the School of Public Affairs at Baruch College provided numerous helpful comments and suggestions. We thank, among many, Paul Kleindorfer, Andrew Kleit, and Sanders Korenman. Eric Hirst, Carlos Ruffin, and Sanford Berg provided very useful help in our understanding of ancillary services, the Chilean electricity experience, and the public power sector in the United States, respectively.

We also want to thank the anonymous outside reviewers, whose careful and thoughtful reading improved this book greatly. Needless to say, the views expressed in *Alternating Currents* need not be shared by any of these people or institutions mentioned above, and errors are solely our responsibility.

A Final Word

One last comment is in order. As we noted, for much of 2000 and the early part of 2001, energy issues, particularly California's crisis, dominated the headlines. That has not been true since September 11, 2001. As noted above, the industry continues to evolve and sometimes makes the headlines as it did during the California situation. Electricity remains a crucial component of our economy, and ensuring reliable access at reasonable prices deserves a prominent slot in the policy agenda. However, the tragic events of that day remind us that we should only be so lucky to be able to return to an era when electricity deregulation is the biggest concern that our nation faces.

TIM BRENNAN, *University of Maryland, Baltimore County and Resources for the Future*

KAREN PALMER, *Resources for the Future*

SALVADOR MARTINEZ, *University of Florida*

About the Authors

Timothy J. Brennan is a professor of policy sciences and economics at the University of Maryland, Baltimore County and a senior fellow with the Quality of the Environment Division at Resources for the Future. His primary research interests are antitrust and regulatory policy, with particular focus on the electricity, telecommunications, and computer software industries. He is one of the authors of *A Shock to the System: Restructuring America's Electricity Industry* (RFF, 1996), and his articles have appeared in journals covering a variety of disciplines, including economics, law, communications, philosophy, and history. In 1996–1997, he served as a senior staff economist for the White House Council of Economic Advisers. He received his Ph.D. in economics from the University of Wisconsin in Madison in 1978.

Karen L. Palmer is a senior fellow in the Quality of the Environment Division at Resources for the Future. Her primary research interests are the economics of environmental regulation and public utility regulation. Her current research focuses largely on electricity regulation and the environmental regulation of the electricity sector. In 1996–1997, she spent six months as a visiting economist at the Office of Economic Policy at the Federal Energy Regulatory Commission. She is one of the authors of *A Shock to the System: Restructuring America's Electricity Industry*, and her published papers have appeared in many journals, including the *RAND Journal of Economics*, the *American Economic Review*, the *Journal of Political Economy*, the *Journal of Public Economics*, and the *Journal of Environmental Economics and Management*. She received her Ph.D. in economics from Boston College in 1990.

Salvador A. Martinez is a Ph.D. candidate in the Department of Economics at the University of Florida. His dissertation research focuses on the economics of solid waste and recycling. His broader interests include public utility regulation, state environmental regulation, and transportation planning. He has performed research for the Bureau of Economic and Business Research and the Public Utility Research Center at the University of Florida.

Tim Brennan dedicates this book
to his long-time companion Judy Boggess
for her unstinting encouragement and
for everything she does
to keep him on the path.

Karen Palmer dedicates this book
to her husband Phil Twomey
and to her children, Josie and Tom,
for providing encouragement,
moral support, hugs, and
other ancillary services during
the writing of this book and always.

Salvador Martinez dedicates this book
to his grandmothers and
to the memory of his grandfathers
for their inspiration, character, and spirit.

Alternating Currents

CHAPTER 1

Issues in Restructuring the Electricity Industry

As the twenty-first century begins, the U.S. electricity business is in the midst of a revolution. An industry that has been dominated by monopoly utility companies, and regulated from top to bottom by the states and the federal government, is now seeing competition and deregulation in the generation and sale of electric power.

These changes are both facilitating and being facilitated by the wider role that new, independent generators are playing in the electricity sector. In addition, promoting competition has become associated with rules, regulations, institutions, and in some cases divestitures that are designed to ensure that power markets operate efficiently and competitively. For that reason, the process of enacting and implementing laws and policies to bring more competition to electric power markets has come to be known as *restructuring.*

Opening markets to competition generally gives firms better incentives to control costs and introduce innovations. Competition among firms means that the benefits of such efforts get passed down to consumers as better service at lower prices. The hope underlying restructuring is that the $200 billion electricity sector will see benefits comparable to those achieved from opening other industries to the market.

Although the term *restructuring* may be unique to electricity, the sometimes painful process of undergoing a transformation from regulation to competition is not. Electricity is one of the last of a series of industries in which market forces have been introduced to take over the duties of choosing product characteristics, determining supplies, and setting prices. The U.S. economy has coped with, and in large degree profited from, similar upheavals in the banking, transportation, and telecommunications sectors. Like those other sectors, electricity is itself a major industry, as well as forming part of the country's economic backbone. Thus, much of the experience with opening other sectors is useful in designing policies to restructure the electricity industry.

Yet electricity has some unusual attributes that present thorny problems for those charged with expanding competition. To explain these problems, we provide in Part I some background on the electricity industry, including the technology for producing and delivering power, the history of policy and regulation directed toward it, and recent experience with restructuring both in the United States and internationally. The dominant episode in restructuring so far is the California crisis, beginning in the summer of 2000. We assess the events and the long list of possible causes of the disaster in Chapter 5.

We then turn to the major questions facing policymakers as they deal with deciding whether, when, and how to implement restructuring. These include matters such as industry structure, future regulation, maintaining system integrity and reliability, promoting competition, and protecting the environment. The "issue chapters" that make up Part II can be read on their own for readers interested in particulars, although we provide links to other chapters to highlight the frequent and significant occasions when one policy issue affects and is affected by others.

We conclude with some background on other political controversies associated with electricity restructuring, including disputes between households and industrial users, between consumers and marketers, and between states where electricity is cheap and those where it is expensive. Although these issues are important, they can divert attention from the most crucial question, which in our view is whether the United States can reap the benefits of competition in electricity while preserving the reliability upon which the nation depends.

Background: Technology, History, and Recent Experience

Evaluating whether and how to bring competition to the electricity industry requires first some understanding of how electricity gets from power plants to the outlets in the wall. The path entails generation of the power, transmission over long distances, and local distribution into the home. Electricity can be generated using different technologies, with different properties regarding how quickly they can be brought online to meet power needs and how much they pollute the air when they operate. The transmission and distribution sectors are subject to important economic conditions that are likely to preclude competition in those sectors in the near future. Transmission, in particular, also exhibits some important technical characteristics that make pricing and management particularly complex when generators want to send more power through the lines than the lines were designed to carry.

Seeing why the move to expand competition in the industry is happening now also requires some understanding of how we regulate the power industry. The industry has a long history of policy attention, dating back to state regulation of electricity rates starting around the beginning of the twentieth century. Major federal laws include the Public Utility Holding Company Act of 1935 and the Public Utility Regulatory Policies Act of 1978 (PURPA). The earlier law set the rules for organizing utility companies, while the later statute opened wholesale electricity markets to selected new power producers, primarily as a means of reducing use of fossil fuels.

The 1992 Energy Policy Act, followed by the Federal Energy Regulatory Commission's (FERC's) Orders 888 and 889 in the summer of 1996, expanded PURPA's initiative by allowing all generation companies to use interstate transmission lines.

About three and half years later, FERC issued Order 2000, the successor to Orders 888 and 889, which required each utility to either commit their transmission assets to an independent "regional transmission operator" or explain why it was not doing so. These federal rules have set the stage for the states, which have the primary role in deciding if, when and how to bring competition to the *retail* buyers—households, offices, shops, and factories. As we write, the federal government is pondering the extent to which it should become involved in this effort.

Unlike other movements to introduce competition into formerly regulated industries, such as telecommunications, banking, and transportation, the United States is not among the first to open its markets. Experiences of states and other countries that have established retail electricity markets provide important lessons for federal and state policymakers still evaluating these issues, and also for the consumers and businesses that depend on how well this industry and its regulators do their jobs. These domestic and international experiences offer lessons to teach others just setting out on the road to electricity competition, or deciding whether or not to begin the journey.

Foremost among these experiences is the imposition of higher rather than lower prices, accompanied by rolling blackouts and utility bankruptcies, arising in California in the summer of 2000. The specific phenomena that have raised angry voices against restructuring include very high prices during times when electricity is most in demand, usually hot summer afternoons. We review the history of these important developments, identifying 10 possible culprits relating to supply conditions, market design, and market power. Our assessments reflect that the factual issues, financial calamities, political controversies, and regulatory responses remain open, and that these issues may arise in different forms in other states.

Leading Issues

Knowing something about generation and delivery technology, regulatory history, and recent restructuring experience allows us to provide some answers to the 11 most compelling questions facing electricity policymakers today. We introduce these questions here; each of them is analyzed in detail in its own chapter in Part II of this book.

1. How—and why—do we draw a line between regulated and competitive sectors of the electricity industry? In many industries where we as a society have elected to replace regulation with competition, such as trucking or banking, much if not most of the industry has been largely freed from continued regulatory oversight. But in some sectors—telecommunications, for example—the process of deregulation has been only partial, with continued regulation of some segments. If we could just deregulate and walk away, the policy task would be much simpler.

When deregulation is partial, it becomes more complicated. Policymakers need to decide where regulation should end and deregulation ought to begin, how best to continue regulation where necessary, and how to manage the new problems of controlling how the regulated and newly competitive portions of the industry relate to one another. We look at why electricity is one of those industries where parts of it—the wires that carry electricity from the power plant to the home, office, and factory—will remain regulated. Paradoxically, regulating less of the electricity industry could make regulation itself more complex. Perhaps technologies in the offing allowing generation at the user's location could reduce the need to regulate the industry altogether.

2. Should the same companies own and control both regulated "wires" and competitive generation? Much of the effort to eliminate price regulation during the past couple of decades in the United States has involved more or less complete deregulation throughout the entire sector. Electricity is an exception. Policymakers are opening power markets to competition, but local distribution and long-distance transmission are unlikely to be deregulated any time soon.

Electricity is not the first deregulated industry to be split into regulated and competitive sectors. The telecommunications industry has seen a transformation from one in which regulation set all prices, to one in which markets for telephone equipment and long-distance service have been opened while local telephone service has, until very recently, been treated as a regulated monopoly. U.S. experience with that sector provided the lesson that letting the regulated monopoly continue to operate in competitive markets could subvert competition in a number of ways. The regulated firm might put one firm ahead of others in the queue for getting access to the regulated service or have the customers of its regulated services bear the costs of its competitive ventures.

Ultimately, these concerns led in 1984 to the draconian solution of keeping most regulated local telephone companies out of the long-distance business, a restriction only slowly changing since the Telecommunications Act of 1996. In electricity, state and federal policymakers must wrestle with a similar decision: Should regulated wire monopolies be prevented from owning generation facilities? Can other operational institutions and rules ensure that transmission and distribution monopolies promote competition without forcing utilities to divest all of their generators? The widespread use of the term *restructuring* to describe the introduction of power competition into the electricity industry illustrates just how fundamental these concerns are.

3. Because we have to regulate prices for the wires, how do we set their rates? As we will see in Chapter 6, the wires segments of the electricity sector—transmission and distribution—will continue to be regulated. Regulation, of course, is not a new issue in this industry, but most regulation before restructuring has been devoted to setting the electricity rates that users pay. With restructuring, power prices will be set by the market, with prices users pay directly or indirectly including those power prices plus the regulated charges for delivering electricity from the generator to the customers' premises. We discuss first methods for setting rates for transmission and distribution, describing both traditional "rate-of-return" regulation and new "incentive-based" methods that could lead to lower costs and more efficient operation.

Although these principles apply to both distribution and transmission, the latter presents difficult problems. A generator may have to go through lines owned by a number of different utilities in a number of different states. Consequently, policymakers must consider whether transmission prices should be set by broad geographical regions and independent of distance, or include charges that increase with distance or the number of times the path crosses a state line or uses a different utility's facilities. An even more complex set of questions related to pricing and incentives is associated with the possibility that transmission lines may be congested.

4. What do we need to do to keep electricity markets competitive? The belief that opening retail markets will lead to lower prices and better service for households, offices, and industrial users is predicated on the belief that electricity generation markets will be competitive. Such markets may fail to be competitive if only one or a small number of firms supply power to a particular area, or if the power producers agree among themselves not to compete. As we observe an industry in flux, with numerous mergers, divestitures, entrants, and volatile prices, how to ensure competition becomes an ever more pressing question.

The antitrust laws are the main legal means of ensuring that competitive markets remain that way. Because those laws are not designed to control markets, such as electric power, where monopolies arose as a matter of prior regulation, a first policy step in some states could be to require divestiture of power plants in such a way as to increase the number of independent competitors.

One concern, presented by the California electricity crisis, is that generators may unilaterally find it profitable to withhold output to raise prices, even when the markets appear competitive by conventional structural indicators. Such concerns have been behind calls for temporary federal caps on wholesale prices. Evidence supporting assertions that market power is being exercised needs to be handled with care. Modifying the operations of electricity markets or programs to make consumers more sensitive to prices (e.g., installing "real-time meters") may reduce the incentive for anticompetitive withholding. If not, then wholesale price caps, particularly during peak periods, could become a permanent feature of "deregulated" wholesale electricity markets.

Mergers among firms that compete could give the firms the ability to raise prices on their own, facilitate collusion among all the competitors, or make competition less intense. Deciding whether to block a merger requires understanding who competes with whom, how competitive the market might be, and who might enter if the price goes up. In some cases, mergers between a generation company and gas companies could cause problems if the gas company is a primary supplier to that generation company's competitors. Finally, although the industry is in transition, merger evaluation could be so speculative that antitrust authorities may have too hard a time proving that a merger may be harmful.

5. Who should be responsible for keeping loads balanced and dispatching power? Electricity stands out in that, unlike virtually every other commodity, disaster can strike unless producers supply exactly the amount that people want to buy at any given time. Keeping power production and use in line—load balancing—will require the active involvement of generators, transmission companies, local distributors, and customers, as well as the regulators who oversee the industry. In implementing electricity restructuring, policymakers must consider how to guarantee the provision of ancillary services needed to keep loads balanced on a minute-by-minute basis and provide emergency power when generators or transmission lines unexpectedly fail or demand is unexpectedly great.

A first question is whether each generation company should be responsible for keeping its own power supply in balance with its own customers' desires. Because failure to meet power demands causes a breakdown of the system as a whole and not just a blackout to that company's customers, letting the market take care of it may not suffice. Generators may need to meet standards for maintaining power and having reserves available, or they may need to be held liable when their inability to meet demand brings down the larger grid. If those measures prove inadequate, distribution and transmission companies may need to take on the responsibility of providing ancillary services and holding power in reserve.

Involving grid operators in the business of maintaining loads has led many states to also involve them in the overall management of power markets, through taking bids from producers and users and dispatching generators as needed. The grid need not be involved in this aspect to control generation costs; the electricity market, like any other, can handle that through letting generators compete for customers. But whether such a market is compatible with keeping loads balanced and systems secure is perhaps the crucial question facing electricity policymakers.

6. As utilities compete, how can we ensure reliability? The U.S. electric power system has had a strong record of uninterrupted service made possible through the cooperative efforts of the utilities that are linked together on the three major U.S. transmission grids. As the electric power industry becomes more competitive, this voluntary approach to ensuring reliability is threatened at the same time that the transmission system is facing greater stress from more intensive use.

Reliability can be classified in terms of adequacy and security. In a competitive world, the market is expected largely to handle generation adequacy. However, transmission and distribution adequacy will still be subject to regulatory oversight. The security of the power system will remain a responsibility of centralized system operators due to the large spillovers associated with failure of generators or transmission lines.

Restructuring poses challenges for the reliability of both the distribution system and the bulk power transmission system. However, the threats to its integrity and the consequences of failures are greater for the transmission system than for a local distribution grid. To maintain the security of the bulk power transmission grid, power control area operators and security coordinators may need to interfere with the commercial transactions on the electricity grid. However, distinguishing an action taken to protect system security from an action taken for other reasons, perhaps anticompetitive ones, may be difficult.

Given the potential threats to reliability posed by electricity restructuring, legislators and energy regulators should develop a strategy to protect system reliability as they design and implement policies that set the course for electricity markets in the future. Such a reliability strategy is likely to include expanding the role of industry reliability councils and federal regulators in overseeing reliability and increased use of incentives to promote efficient use of the transmission and distribution systems.

7. Should the states or the federal government control the course of retail electricity competition? So far, state governments have been the key actors in developing and implementing policies to encourage retail electricity competition.

A policy question has been whether states are acting quickly enough, or whether the federal government should step in to encourage or force them to open markets by a particular time.

Keeping control with the states allows the nation as a whole to learn from what worked in one place and what did not work so well someplace else. One size may not fit all, in that the benefits of opening markets may be considerably greater in some states than others. In addition, to impose a federal solution could cause needless and costly difficulties in trying to amend or reverse the delicately balanced solutions achieved by states that are moving ahead in opening retail markets.

However, a presumption that state actions might reflect a proper balance of interests is less convincing when that state's decisions have effects that go across their boundaries. When interstate effects are significant, the federal government can help improve policies by serving as a venue where all affected parties have a say. Specific areas in which the federal government can play an effective role include reforming existing federal laws that may inhibit competition, regulating interstate transmission grid prices and operation, enhancing market liquidity, enforcing antitrust and environmental laws, and coordinating commercial standards and practices. Also, states themselves may be able to negotiate solutions and set up regional authorities to manage issues that affect an entire region but not the nation as a whole.

8. What should be the role of public power after restructuring? Unlike most of the other industries that have made the transition from regulation to competition, the electricity sector has a substantial nonprofit component. Roughly 25% of all retail electricity sales in the United States comes from publicly or cooperatively owned utilities. The combination of privately and publicly owned utilities (at local, state, and federal levels of government) operating under different objectives and rules greatly complicates the task of restructuring the electric power industry.

The debate over bringing competition to electricity generation and retail sales markets has highlighted several differences between publicly and cooperatively owned utilities and investor-owned utilities. These differences can be categorized into three types: financial, regulatory, and scope of activities. The first category refers to the special privileges granted to public utilities and cooperatives, which include preferential access to low-cost hydroelectric power produced at federally owned facilities, the ability to issue tax-exempt debt, and exemption from income tax payments. The second category refers to the exclusion of publicly owned and cooperatively owned utilities from the state and federal regulatory structures governing investor-owned utilities and other regulations that could limit the ability of publicly and cooperatively owned utilities to participate in regional transmission organizations. The third category refers to the fact that many federally owned hydroelectric generation facilities have multiple purposes, such as flood or navigational control in addition to electricity production. If publicly owned and cooperative utilities are going to compete in electricity markets, policymakers must address these differences as they seek to make those markets fair and truly competitive.

How public power will evolve in this era of competition remains an open question to be decided at different levels of government. The federal government must define the new role for the federal power marketing authorities and for the Tennessee Valley Authority. Decisions about whether or not municipal utilities or rural cooperatives will continue to hold an exclusive franchise for retail electricity sales are best made at the local level.

9. Will it cost utilities to adapt to competition and, if so, who should pay? Before the California electricity crisis, perhaps the most highly charged (pun intended) issue associated with opening electricity markets to competition was whether and how to compensate utilities for capital expenses they incurred during the regulatory era. If competition brings about lower prices, as its advocates would hope, utilities fear that they would not make enough money to recover some of these costs—hence that they would be stranded. The primary sources of stranded costs, once estimated at upward of $135 billion, are associated with nuclear power plants and long-term contracts to purchase renewable and cogenerated power under the PURPA (see Chapter 3).

Utility advocates argue that a "regulatory compact" implicitly guaranteed cost recovery as part of the utilities' obligations to provide service. Those opposed to stranded cost recovery allege that utilities should not be rewarded for unwise investments and that forcing consumers to pay for stranded costs will thwart the objective of reducing electricity rates. In principle, deciding who is right should turn on a determination of whether regulators or utilities were in the best position to foresee restructuring, and which of them were better able to adapt to the prospect of competition.

As a practical matter, stranded cost recovery has generally been part of the package necessary to build sufficient political support to implement opening retail markets. In addition, the federal government supports stranded cost recovery—perhaps not incidentally because the federal government is itself exposed by virtue of its ownership of electricity generation in the Tennessee Valley and the Pacific Northwest (see Chapter 13). If *stranded costs* (i.e., investments that a utility cannot recoup because competitive prices are low) are recovered through surcharges on electricity purchases, it is important to devise methods that preserve competitive neutrality (i.e., do not introduce fees that create artificial cost advantages for either incumbent utilities or new merchant generators). Designing such a recovery system may be easier said than done.

10. What are the implications of restructuring for environmental protection? Electricity generation is a major source of air pollution in the United States. In the midst of searching for new ways to reduce air pollution in general, environmental regulators and other policymakers are eager to understand how increased competition in electricity markets is likely to affect the size of that sector's contribution to different air pollution problems. The effect of electricity restructuring on the amount of air pollution emitted by the electricity-generating sector will depend on three key factors: how competition affects the size of the market for electricity, how competition changes the mix of technologies used to generate electricity, and the form of existing environmental regulations governing electricity generators. In general, competition could lead to greater emissions of those pollutants, which are not subject to strict caps, such as carbon dioxide, unless additional provisions are made to make the use of renewables and cleaner technologies more attractive.

Opening electricity markets is also likely to affect the performance of environmental regulation. Competition will likely limit voluntary actions to reduce emissions. At the same time, it will enhance incentives for electricity generators to take advantage of emissions trading. Environmental regulation and plant-siting requirements could limit incentives for investment in new generation by potential competitors, with adverse effects on market performance. However, the magnitude of these effects is largely unknown and may be dwarfed by the investment-reducing effects of general uncertainty about the future of electricity restructuring.

11. What happens to utility-funded "public benefit" programs in a competitive electricity market? Regulated electric utilities historically have performed several public service functions in addition to selling electricity. These activities range from offering rebates to consumers who purchase energy-efficient appliances—so-called demand-side management programs—to funding industrywide research and development of more efficient generating technologies. All have been made possible by the fact that regulators have, for the most part, allowed the regulated utilities to recover the costs of these activities in the prices that they charge electricity consumers.

In the newly competitive environment, utilities face greater pressures to reduce costs and therefore are reducing discretionary spending on optional activities, such as public purpose programs, that do not contribute directly to profits. At the same time, competition brings with it important changes in the incentives facing electricity suppliers and consumers that could eliminate or reduce the need for certain public purpose programs or require change in the means of provision to make them consistent with the reality of a competitive electricity market.

Public purpose activities traditionally funded by electric utilities on the policy agenda include the programs mentioned above as well as promoting renewable energy and protecting low-income users. Competition may affect the sustainability and implementation of these programs. However, competition also can change the justification for public policies to promote these activities. New and proposed policies to promote public purpose programs may better suit a more and more open electricity market.

Reliability in a Competitive Environment: The Bottom Line

In addition to the fallout from the California crisis and the policy issues outlined above, legislators and regulators dealing with restructuring must handle several other political "hot potatoes." Some complain that big industrial users are benefiting at the expense of households. Consumer advocates want to ensure that residential customers possess sufficient and useful data on prices and environmental effects to make informed choices among different electricity service offerings. Citizens of states where electricity is currently cheap may fear that they will now have to pay higher prices if their suppliers can now sell in "high-cost" states that open their markets to new competition.

These concerns are all significant, but they are the kinds of problems that we as a society have addressed before in dealing with deregulation in other industries. The matters that should continue to hold our attention here are those affected by the factors that make electricity different than any other good or service. In our view, the primary differences arise from the combination of three characteristics of electricity: It is *critical* to the economy; it is *technically exacting,* in that supply needs to equal demand at all times; and it is *interrelated,* in the sense that one firm's inability to serve its customers could bring down the entire network.

Together, these factors suggest that the make-or-break issue in electricity restructuring is whether the kind of cooperation that is necessary to maintain reliability in an interrelated, exacting industry is compatible with the degree of competition that is necessary to bring about greater efficiency and lower prices. The importance of electricity to the U.S. economy means that this issue deserves the attention of the public and those elected and appointed to serve it.

PART I

How the Industry Got Here

CHAPTER

Understanding the Electricity Industry

Before it can be decided whether and how to bring competition to the electricity industry, it is important first to gain some understanding of how electricity is produced and how it gets from power plants to the outlets in the wall. The cost and environmental consequences of producing electricity essentially depend on the mix of technologies and fuels used to generate it. Making sure that electricity is available when households and businesses need it requires a complex web of services that complicates the process of deregulation.

Changes in the composition of the industry itself, and in the technologies used to generate electricity, are among the forces pushing the industry toward competition. Some segments of that path are ripe for competition, but others will continue to need regulation. Where competition is allowed, it is creating both opportunities for new entry and incentives for consolidation among some existing market participants.

From Plant to User

The electric power industry comprises four major functions:

- *Generation*: the production of electricity from other energy sources, typically burning of fossil fuels (coal, oil, natural gas), nuclear fission, falling water, or wind.
- *Transmission*: the transportation of electricity over long distances at high voltages, typically from generators to local utility companies.
- *Distribution*: the transformation of high-voltage electricity to lower voltages and the delivery of that power to users for lighting, heating, air conditioning, appliances, and other personal and commercial uses.
- *Marketing*: the advertising, selling, and billing for electricity use.

Traditionally, an electricity consumer has paid one regulated price for electricity to a single *vertically integrated* utility responsible for all four of these functions. (A utility is vertically integrated when it controls two or more parts of the chain of production, e.g., generators and transmission lines. We will discuss vertical—and horizontal—aspects of the electricity industry in more detail in later chapters.) Policies to open markets have led to new competitors in generation and marketing, with a restructuring of the industry away from the regulated, single-provider model.

As we will see in this chapter, electric power can be generated in many different ways. Whatever the generation technology, even the largest units are not so great as to preclude the possibility that many of them could potentially supply power to any customer, at least as long as the transmission lines that deliver power to that customer are not congested (see below). The effort to bring about competition in electricity is, at its essence, based on the recognition that generation and marketing are not inherently monopoly businesses.

Generation

Technologies and Fuels. Electricity is generated using a variety of different technologies and fuels.

■ *Fossil-fired steam turbine plants.* About 59% of all the electricity supplied by the U.S. electric power industry comes from steam turbines fired by fossil fuel. Most of this electricity is produced using coal as the fuel. In these plants, coal or another fossil fuel is burned in a boiler to produce high-pressure steam. The steam turns the blades of the turbine, which then rotate an electric generator, which converts the physical energy into an electrical current.

The average coal-fired steam turbine has a capacity of roughly 250 megawatts (MW)—roughly the amount of power needed to supply a town of 60,000 homes (see box on next page). As a general rule, coal-fired steam plants, which contain multiple steam turbines, require a capacity of 300–600 MW to produce at an output level that roughly minimizes the average cost of producing a kilowatt-hour (kWh) of electricity. Construction costs for a typical new coal-fired steam plant are now about $1,100 a kilowatt. The operating costs depend on the cost of fuel and typically range from less than 2 cents to a little more than 3 cents per kWh, depending on the age of the plant and the types of environmental controls it is required to have.

■ *Nuclear plants.* Nuclear power plants generate about 20% of the nation's electricity. This technology uses nuclear fission, the splitting of uranium atoms, to create the heat that is used to produce steam for the power plant turbines. One pound of uranium can generate more than 3 million times the energy produced by a pound of coal. Although no new nuclear plants have been ordered since 1979, nuclear plants will continue to generate a substantial amount of electricity for several years into the future.

Nuclear plants are among the most expensive types of plants to build. Nuclear plants also can have high maintenance costs, which are due in part to the costs of complying with safety regulations imposed by the Nuclear Regulatory Commission. An average nuclear plant is approximately 1,000 MW in size, which makes it possible to spread the construction and maintenance costs over a large amount of electricity production. Utilities that operate nuclear plants are also required to collect money to put into a trust fund to cover the future cost of safely removing the plant from operation and reducing residual radioactivity, so-called decommissioning.

The costs of operating these plants are generally quite low, in the neighborhood of 1–2 cents per kWh.

■ *Gas turbines.* In a gas turbine, or combustion turbine, hot gases created from the combustion of natural gas or distillate fuel oil pass directly through the turbines that spin the electric generator. To create a higher level of efficiency, some of these turbines use exhaust from the combustion to preheat the air that is mixed with the fuel under high pressure before combustion. Gas turbine capacity accounts for roughly 10% of total generating capacity but 2% of total commercial electricity generation in the United States.

This high ratio of capacity to generation reflects the fact that gas turbines are generally used to meet peak demands. Gas turbines have relatively low capital costs, but substantially higher operating costs than steam turbines. The average gas combustion turbine has a capacity of roughly 35 MW, but the size varies from the larger, permanent gas turbines to the newer and smaller aeroderivative gas turbines developed from research investments in jet engine technology. The typical new gas turbine has a construction cost of roughly $260 per kilowatt and operating costs that are 30% to 50% higher than a typical coal plant, depending on the relative prices of natural gas and coal.

■ *Combined-cycle gas turbines.* Combined-cycle gas turbine (CCGT) generators use both combustion turbine and steam turbine technologies. CCGT units use the exhaust heat from the gas turbines to create steam for the steam turbines. About two-thirds of the power is generated by the gas turbine, whereas a third is produced by the steam turbine. These units are compact, reliable, and versatile, and require short installation times. *Energy efficiency,* or the amount of electricity produced per unit of fuel, for a CCGT can be as much as 70% higher than a typical coal plant and 40% higher than a gas turbine plant. CCGT plants produce about 8% of all electricity.

New CCGT plants can now be constructed at a cost of $450 per kilowatt, less than half the cost of a coal-fired steam plant. The fuel costs for a 300-MW CCGT plant made up 85% of total operating costs (fuel, operating, and maintenance) in 1998, whereas the fuel costs for a coal plant of the same size made up 78% of total

Measuring Power and Energy

The two most important measures of generation involve the total energy produced or used over time, and power, which we can think of as the ability to produce or use energy at any given instant. The standard measure of power is the *watt,* with *kilowatts, megawatts,* and *gigawatts* referring to watts measured by the thousands, millions, and billions.

Energy can be thought of as the amount of power used during a given period of time. The primary unit of measurement in electricity is the *watt-hour* (Wh), which uses 1 watt of power for 1 hour. One can calculate watt-hours by multiplying the amount of power used by the amount of time it is used. For example, keeping a 100-watt lightbulb on for 10 hours uses 1,000 watt-hours, or 1 *kilowatt-hour* (kWh). As with power, one can measure energy by the *megawatt-hour* (MWh) or *gigawatt-hour* (gWh), that is, in millions and billions of watt-hours.

Generally speaking, energy is the measure of the output we buy and sell in electricity markets. Prices vary considerably by region of the country, season of the year, time of day, and type of customer (industrial, commercial, or residential), but on average, electricity costs about 6–7 cents per kWh, or $60–70 per MWh in the United States.

Power, however, represents capacity to produce electricity or capacity to use it. One would describe a generator in terms of the power it can produce, typically measured in kilowatts or megawatts. Watts also measure the capacity of an appliance to use electricity by the power it drains from the system at any instant (e.g., a 100-watt bulb, or a 10-kilowatt air conditioner).

One watt is also the amount of power necessary to send 1 *ampere* (or *amp*) of current through a line with 1 *volt* of electromotive force. A transformer is a device that can increase (or decrease) voltage by a given factor, and simultaneously decrease (increase) current by the same factor, which keeps overall power constant.

operating costs. As a result, new CCGT plants are competitive with new coal-fired plants and they have much lower emissions. Combined-cycle plants are being built at a rapid rate, with total capacity predicted to grow by roughly 60% during the next four years.

■ *Internal combustion generators.* Using diesel fuel or natural gas, internal combustion generators operate in a similar fashion to the engines in diesel-electric locomotives by using an explosion to drive a generator. These units are compact and able to start up and shut down almost instantaneously by remote operation. Most are from 1 to 3 MW in size. Only a very small fraction of total U.S. generation comes from internal combustion generators.

■ *Hydroelectric power.* About 8% of U.S. electricity comes from hydroelectric generation (or hydropower), the process of using the flow of water to spin a turbine. The two primary methods for creating hydropower are dams that create reservoirs of water, which then flow through the turbines, or plants on rivers that use the natural flow of the current to propel the turbines. A third type of hydroelectric plant, a pumped storage facility, uses electricity during low demand periods to pump water to elevated reservoirs. The water then can be released through hydraulic turbine generators when demand for electricity is high.

Most of the expenses of producing hydropower are the fixed costs of constructing and maintaining the dam and of meeting environmental requirements, such as building passageways for migrating fish. The fuel costs of hydropower facilities are zero, and the operating costs thus are close to zero as well.

■ *Nonhydroelectric renewables. Renewable* power comes from energy sources that are generally viewed as both friendly to the environment and essentially inexhaustible. Falling into the "renewable" category are wind, solar energy, geothermal energy, and biomass combustion. Wind power is derived from windmills connected to an electrical generator. Solar power is produced with photovoltaic (solar) semiconductor cells that create electricity directly, or parabolic mirrors that direct the sun's rays to heat water and drive steam turbines. Geothermal power uses steam obtained directly from the earth from water heated by volcanic or other geological processes. Biomass power involves burning municipal solid waste, wood, agricultural products, and landfill gas to create electricity, usually with steam-powered technology.

Together, renewable sources other than hydropower contribute roughly 2% of total electricity generation (see Figure 2-1). With the exception of wind and biomass, most of these technologies are not generally competitive with fossil and hydroelectric power, except in special circumstances. One disadvantage of both wind and solar technologies is that it is difficult to control when they generate; windmills generate electricity only when the wind is blowing, and solar power is generated only when the sun is shining. However, many policymakers and environmental advocates have proposed regulations to triple renewable fuel use in the next 10 years.

Dispatching Generators

Demand for electricity fluctuates depending on the time of day, season of the year, weather, and general level of economic activity. Some amount of electricity demand

Ancillary Services

Keeping a distribution network, control area, or transmission grid balanced and running nominally requires electric power. *Ancillary services* are the power-related functions necessary to keep the system working and reliable. Chapter 9 discusses these in more detail. Examples include:

- *Regulation*: maintaining central control over generators to adjust power instantaneously to deal with momentary power surges and reductions in demand. Hydroelectric power is one of the generation technologies most suited for this service in competitive markets because it has the ability to quickly change generation levels while keeping variable costs low.

- *Load following*: adjusting generation to adapt to predictable hour-to-hour and daily variation in demand. Load following requires greater capacity but does not require the same flexibility to respond to load changes as regulation generators. Also, the generators do not require the specialized controls to dispatch, because manual dispatching can occur.

- *Reserves*: providing power in response to unexpected generator or transmission system failure. Reserves can be subdivided into generation that can be called on very shortly (i.e., in 10 minutes) and power that can be called on in an hour. *Spinning reserves* refer to the former, in that quickly available replacement power requires that the generator be already on, but not already providing power at capacity. Some of the peaking units, such as pumped hydroelectric power and combustion turbines, have this ability to immediately provide capacity. *Quick-start* generators can also provide reserve power in a short time. *Nonspinning reserves* can come from generators not already on, but that can be ramped up within an hour.

Additional ancillary services, such as voltage control, are important for the operation and reliability of transmission but are associated with more complex issues of measurement, transmission or generator responsibility, location, pricing, and costs.

Traditionally, ancillary services have been supplied by vertically integrated utilities with the cost included in the price of electricity. Some analysts suggest that many of these services can be provided competitively. Chapter 10 also investigates how the design and feasibility of such markets depend on whether the responsibility for load balancing should rest with generators or with transmission operators.

Transmission

Technology and Cost. Transmission allows electricity to be transported at high voltages and long distances from generation plants to localized users of electricity. To minimize energy loss in sending electricity through power lines, transformers decrease the current coming out of a generator while increasing, or "stepping up," the voltage. Heavy-duty wires carry the electricity along high towers and underground. Once the electricity reaches its destination, transmission substations use voltage transformers to "step down" the voltage for local distribution systems or for direct industrial use.

Most of the cost of transmitting electricity is in the constructing and maintenance of transmission systems. However, there are some costs related to the amount of power delivered. Even at high voltages, some of the electric power transmitted across the transmission grid is lost as heat because of the resistance in the wires. Consequently, the amount of electrical power injected into the power grid at a delivery point will be greater than the amount of electricity taken off the power grid at a community location. The amount of energy lost in a typical transmission and distribution system can be as high as 10%.

A more complicated cost has to do with the *congestion* of transmission lines. As the power delivered through a line increases, so too does the heat. If a transmission line were to carry power beyond some amount, thermal breakdown could occur. This thermal limit for a transmission line determines its capacity. To keep lines from exceeding their thermal limits, transmission system operators may have to restrict the output of certain generators and increase the output of others at different locations. Developing the proper markets and incentives to get power suppliers to incorporate congestion-based transmission constraints into their planning and operations, and to get transmission companies to expand their facilities in the most cost-effective way, remains one of the most difficult problems facing industry regulators and policymakers.

Loops and Grids. Exacerbating this management problem for transmission operators is a phenomenon known as *loop flow*. The flow of electric power across a transmission network cannot economically be directed along a particular line. Instead, electric power flows across all available transmission lines according to the physical properties of electricity and the capacities of the different lines on the system. Loop flow is the name given to the characteristic of electricity that it takes all available routes to get from one point to another.

The presence of loop flow means that any one generator's or power customer's use of the transmission grid affects the amount of transmission capacity available to all users, particularly when parts of the system are congested. Moreover, the transmission systems owned by different utilities are typically interconnected to facilitate power exchanges and sharing among them. This interconnection means that the loop flow effects permeate not just one company's lines, but the entire system, or *grid*.

Three transmission grids or *interties* cover the United States:

- The Eastern Interconnected System, which serves the majority of the Eastern United States and parts of Canada.
- The Western Interconnected System, which serves the west side of the Rocky Mountain area, part of Texas, and parts of Western Canada and Mexico.
- The Texas Interconnected System, which serves most of Texas and is connected to the other two U.S. grids.

Within these three electric grids are about 150 *control areas*, which are electrical geographic areas with control operators that balance electric load while maintaining reliability. More than 100 control areas are located in the Eastern interconnection, but fewer than 12 are in the Texas interconnection.

Because of spillovers from loop flow and interconnection, the transmission grid requires coordinated, centralized management, despite separate ownership. Chapters 6 and 8 discuss how these factors imply that electricity transmission is unlikely to be a competitive business.

To improve efficiency in the system, in some regions, largely in the East, companies joined together to form *power pools* to coordinate generation and use of the transmission system by members of the power pool. Non–pool members have had difficulties gaining access to transmission facilities owned by pool members, such as facing complicated use rules and financial terms. To protect against a utility or pool using its control over transmission lines to keep unaffiliated generators from competing effectively, some of the power pools have evolved into nonprofit organizations known as independent system operators and regional transmission organizations. Policy issues in the design and oversight of these entities are the subjects of Chapter 7.

At the national level, utilities formed the North American Electric Reliability Council (NERC) in 1968 to coordinate efforts to improve the reliability of the U.S. electricity supply. NERC came about as a response to the blackout of November 9, 1965, which affected about 30 million people in the U.S. Northeast and Ontario in Canada. NERC and its 10 regional reliability councils are responsible for establishing the standards and guidelines to ensure coordination among utilities exchanging electricity on the power grids, the interconnections that operate synchronous networks. Chapter 11 contains more on how restructuring may affect the reliability of the grid as utilities that have to cooperate now compete as well.

Distribution and Marketing

Distribution is the process of moving electricity from the high-voltage transmission grid to lower voltages and delivering it to businesses, factories, government, and households. The delivery infrastructure requires a large initial investment, similar to that for the transmission network. The cost of serving additional customers is low relative to the initial costs of building distribution substations, primary distribution lines, and main switches. Largely for this reason, distribution utilities will remain monopolies in their service area, and continue to be regulated by state regulators, for the foreseeable future.

The process of providing the electric energy to ultimate customers is *retail marketing* or *retail sale*. Marketing as a separate function from distribution is a new phenomenon resulting from the introduction of retail competition in certain states. Traditionally, the retail sales function has been bundled with distribution, in that the local distributor is also responsible for advertising, billing, and customer relations. However, unlike distribution, electricity retailing does not require a substantial fixed investment, and therefore the retail sale of electricity is being separated from distribution and provided competitively as states implement restructuring proposals.

Within two to three years, just under half of all electricity consumers in the United States will be allowed to purchase their electricity from a competitive electricity service provider instead of being required to purchase from their utility distribution company. The specifics of billing responsibilities between utility distribution companies and electricity service providers are continuing to be defined. In the future, this process may include configuring households with new electricity meters that measure real-time energy use so customers can adjust consumption patterns as electricity prices reflect real-time generation prices. It may be that billing is not provided independently and the local distribution company would continue to be responsible for collections, just as local telephone companies were in the early days of long-distance competition.

Industry Participants

We conclude with a brief survey of who buys and sells electricity. Understanding these markets requires that we recall the distinction between wholesaling and retail sale of power. *Retail sale* refers to power sold to those who use the electricity, whereas *wholesaling* refers to power sold to those who resell that power for final use. Traditional vertically integrated utilities have participated in both markets. But this masks a broad diversity in the types of entities engaged in electricity supply. Some are privately owned; some are publicly owned. Some only produce but do not sell power at retail; some sell at retail but do not produce the power themselves.

Buyers

Customers for electricity are typically classified as residential, commercial, and industrial:

- *Residential* consumers include more than 108 million households.
- *Commercial* customers include nonmanufacturing businesses, retail stores, schools, and government offices.
- *Industrial* customers include slightly more than a half-million manufacturing and construction enterprises.

Roughly speaking, each class purchases about a third of the total power sold in the United States.

Sellers

Sellers of electricity fall into the categories portrayed in Figure 2-2.

■ *Investor-owned utilities (IOUs)*. IOUs traditionally have been largely vertically integrated utilities that generate, transmit, and distribute the electricity that they sell to customers living in their exclusive service territory. IOUs are the most important players in electricity markets. IOUs own 71% of the generating capacity owned by utilities and nonutility generators in the United States and are responsible for 74% of all retail sales of electricity.

■ *Publicly owned utilities*. Publicly owned utilities, which account for about 14% of U.S. generating capacity and 15% of electricity sales to final customers, include municipal utilities, public power districts, irrigation districts, and state authorities. The majority of public power systems are located in small towns, but some large cities, including Jacksonville and Los Angeles, have municipal systems. Municipal utilities operate as nonprofit governmental agencies to provide power at cost. Public power systems have access to tax-free financing and preferential access to low-cost federal power. The policy implications of this treatment in an environment where public utilities and IOUs may be competing with each other are discussed in Chapter 13.

■ *Rural electric cooperatives*. Utilities that are cooperatively owned by rural farmers and communities primarily distribute power to residential customers. They are

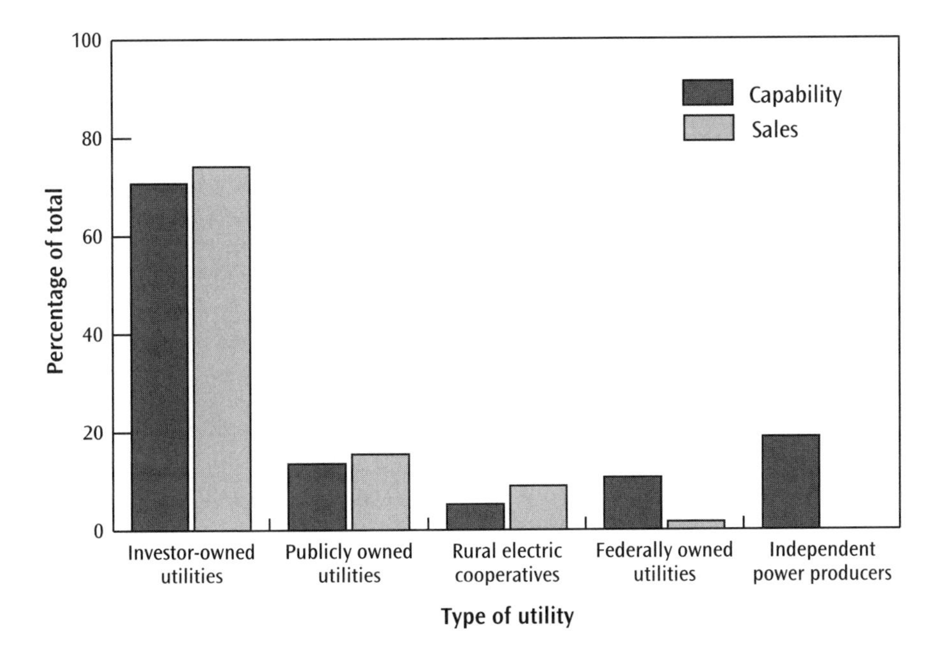

Figure 2-2

Percentage Shares of Total Generating Capability and of Final Sales by Type of Utility, 1999

Sources: Data on generating capability are from U.S. Department of Energy, Energy Information Administration, *Electric Power Annual 2000* (Washington, DC: U.S. Department of Energy, 2001), vol. 1. Data on sales are from U.S. Department of Energy, Energy Information Administration, *Electric Sales and Revenue 1999* (Washington, DC: U.S. Department of Energy, 2000).

found in almost every state, and they serve about 10% of the U.S. population. As is shown in Figure 2-2, the generation capacity of cooperatives is the lowest among the types of electricity providers.

■ *Federally owned utilities.* Federally owned utilities represent a group of five federal wholesale power producers and four federal power marketing administrations. Most of the federally owned wholesale power producers market their power through one of the four administrations, the largest of which is the Bonneville Power Administration. The exception is the Tennessee Valley Authority, the largest federally owned producer of electricity, which markets its own electricity. (See Chapter 13 for more on these administrations.)

■ *Independent power producers.* Independent power producers (IPPs), which are also referred to as nonutility generators, include more than 2,000 generators that are not owned or operated directly by an electric utility with a designated franchise service area. (Nonutility generators may be owned or operated by utility affiliates.) The first nonutility generators included industrial *cogenerators*, that is, factories that used furnaces and other heat-producing technologies to produce electricity for themselves, with some left over for sale to others and independent renewable generators. With the opening of markets in wholesale and retail power, IPPs employing traditional generating technologies are becoming more prominent, in part by buying the generation assets of existing utilities. Collectively, IPPs accounted for about 19% of total U.S. generating capacity in 1999.

This diverse array of participants poses a challenge for policymakers as they seek to create a level playing field for all participants in a more competitive electric power market.

Transformations

As the electric power industry is being restructured, its traditional composition is changing in many ways—especially through new entrants, mergers, and divestitures.

New Entrants

New entrants, including power marketers (see box), power brokers, and retail energy service providers, are playing a more and more important role. Entry by these participants and a host of nonutility generators has helped to swell the ranks of electricity providers in the United States to more than 5,000. With the trend toward increased specialization, a growing portion of the electricity sold to final customers is being traded in wholesale markets, and sometimes traded several times. It is no longer commonplace to have vertically integrated companies perform all the handling of electricity.

Power Marketers

Power marketers are businesses acting as intermediaries in buying and reselling electricity. Although power marketers generally do not own generation, transmission, or distribution facilities, some are affiliated with other companies that own other energy reserves, such as natural gas. In 1999, more than 400 U.S. power marketers were registered with the Federal Energy Regulatory Commission to trade electricity.

Mergers

At the same time, a growing number of investor-owned electric utilities are merging. IOUs claim that these mergers allow for consolidation that reduces the average cost of serving a larger customer base. In addition, recent mergers have brought together electric and gas utilities, which are offering to become complete energy services providers to retail customers. As Figure 2-3 shows, these mergers have led to consolidation in the industry, with the largest IOUs increasing their share of generation capacity. Opponents of these mergers argue that they may reduce competition in electricity markets, leading to higher prices for consumers. Chapter 9 discusses how antitrust enforcers and regulators evaluate mergers to assess such concerns.

Divestiture

At the same time that mergers are becoming more prevalent, many utilities are also starting to get out of the generation part of the electric power business, at least within their own service territories. In several of those states that have gone the farthest toward implementing retail competition, IOUs have been divesting their generation assets, either in response to a legislative or regulatory requirement or by choice. The legislative and regulatory requirements for divestiture are predicated on the desire to separate generation ownership from transmission and distribution ownership to facilitate retail competition. Some IOUs—recognizing the competitive pressures in generation markets and the high capacity levels needed to be successful—are divesting generation assets to concentrate on transmission and distribution. Other IOUs are divesting generation assets to enable their mergers to pass regulatory approval.

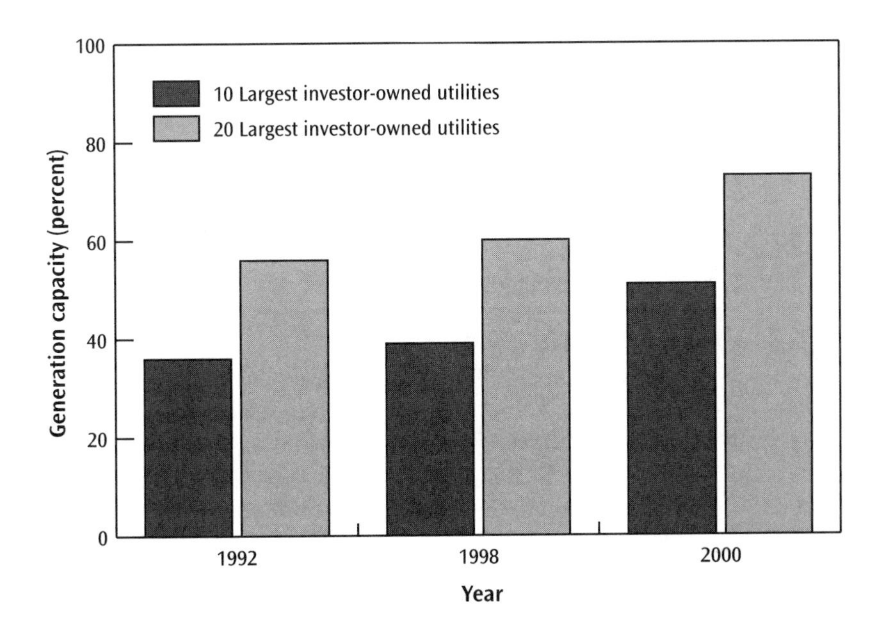

Figure 2-3

Generation Capacity Owned by Investor-Owned Utilities (percent)

Sources: Data on generator capacity are from U.S. Department of Energy, Energy Information Administration, *The Changing Structure of the Electric Power Industry 1999: Mergers and Other Corporate Combinations* (Washington, DC: U.S. Department of Energy, 1999). Data for 2000 are estimates based on pending mergers.

How Did the Industry Get Here?

The current structure, participants, and transformations in the U.S. electricity industry are a result of a movement toward competition. The process of implementing new laws and policies to bring forth more competition in the electricity industry is a restructuring process. Changing the *structure* of the industry has required both federal and state efforts. As a result, *participants* have defined their roles in the industry in response to existing regulations and policies. The *transformations* are a result of a new structure in the electricity industry, which is taking shape as participants seek long-term opportunities with the onset of competition. To explain the current restructuring issues facing policymakers, the next two chapters examine how the power industry has been regulated by federal and state authorities and how restructuring has evolved in some parts of the world.

CHAPTER

3

From Regulation to Competition

For its almost 100 years as a major U.S. industry, the electricity industry has had prices set by regulators rather than the market. The primary justification has been that customers for electricity were unlikely to have the option to choose from competing suppliers in an open market. Most fundamentally, one set of lines suffices to deliver power within a city or town; having multiple providers would be wasteful. The prevailing large size of generators and the belief that generation and distribution were efficiently provided within the same utility led to the belief that the power side of the business was inherently monopolistic as well.

If monopoly is inevitable, electricity prices will predictably be substantially higher than costs. A utility, knowing that it would not lose sales to competitors, would keep raising the price up to the point at which customers would finally and substantially begin to cut back electricity use. Having regulators set prices can prevent utilities from profiting exorbitantly at consumers' expense and can mitigate the overall loss to the economy when high prices discourage electricity use.

Under standard regulatory practice, the government granted to a utility an exclusive franchise to provide electricity within a town, city, or state. The franchise came with general public service obligations. In exchange for the franchise, the utility was allowed to set prices at levels just high enough to cover its operating expenses and to allow its investors to earn a "just and reasonable" return on their investment. Outcomes under this *cost-of-service* regulation may be better than monopoly, but they remain fraught with well-known problems (which will be reviewed in Chapter 8). Consumers may not be helped if the regulated firm is better able than consumers to exercise political influence over how the regulator sets its prices, or if regulation removes incentives to cut costs and innovate.

The promise of restructuring is that it will limit the severity of these pitfalls by letting the market replace regulation in setting prices for generating and marketing power. As we will discuss in more detail in Chapter 6, regulation is likely to remain with us for the transmission and distribution portions of the industry, while generators compete to sell power wholesale and, more and more, at retail to residential, commercial, and industrial customers. But to explain how power markets are evolving and where restructuring can go, we need to look at where we are today and how we got there.

State and Federal Regulators

Today's state public utility commissions (PUCs) typically evolved from railroad commissions established in the post–Civil War years, which gradually extended their scope to cover natural gas, telephone, water, and electric utilities. In the electricity industry, PUCs have been responsible for regulating the distribution and retail sales of electricity. They determined where investor-owned utilities can operate, which facilities they can construct, and at what prices they can sell. PUCs also have the authority to approve or disapprove proposed utility mergers.

The process works a little differently for publicly owned electric companies, such as municipal utilities and rural cooperatives. Because municipal utilities are owned by the local government in the area they serve, and cooperatives are owned by the customers themselves, they may have less incentive than investor-owned utilities (IOUs) to take undue advantage of their monopoly position. Accordingly, in most states, publicly owned utilities and cooperatives set their own prices.

The national analogue to the state public utility commission is the Federal Energy Regulatory Commission (FERC). FERC began as the Federal Power Commission, instituted in 1920 and granted authority over electricity in 1935. In 1977, it received its current name when it was incorporated into the newly created Department of Energy. Along with its authority over interstate gas and oil pipelines, FERC regulates wholesale electric power sales, or so-called sales for resale. It also sets the prices charged for transmission services across states for IOUs, power pools, power exchanges, power marketers, and independent system operators. Like the PUCs, FERC has ability to approve or disapprove mergers. FERC also oversees the licensing of hydroelectric power projects and maintenance activities whose transmission grids meet international borders. Also, FERC regulates natural gas transmission and interstate oil transportation.

FERC is not the only federal agency with an interest in the electricity industry. Because electric utilities are controlled by antitrust laws (see box), they are subject to enforcement actions of the Department of Justice's Antitrust Division and Federal Trade Commission. In addition, under the Public Utility Holding Company Act (see below), the Securities and Exchange Commission retains authority over the internal structure and intracorporate dealings of utility companies. The Nuclear Regulatory Commission oversees the licensing, construction, and safety of nuclear power plants. Other environmental issues fall under the jurisdiction of the Environmental Protection Agency. Last and not least, the Department of Energy designs, coordinates, and implements a variety of policies related to energy security, environmental protection, and the economic performance of the electricity sector.

Antitrust Authority

In addition to federal legislation and regulation, the judicial system has shaped today's electricity industry. A landmark decision is the 1973 Supreme Court case, *Otter Tail Power, Inc. v. United States*. The Court ruled that Otter Tail could not refuse to transmit power from other electricity suppliers to existing or proposed municipal systems using its own wires, commonly referred to as *wheeling*.

Wheeling is essential for a competitive industry because it allows electricity suppliers to reach customers from outside locations. Otter Tail argued that under the Federal Power Act it was not subject to antitrust regulation in deciding to refuse to deal; the Court disagreed. Hence, *Otter Tail* established the authority of antitrust laws over transmission facilities to promote competition. It also established the importance of open access to transmission facilities, contributing to the initiatives to bring competition to wholesale and now retail electricity markets. (For more on antitrust issues in electricity restructuring, see Chapter 9.)

Eighty Years of Federal Legislation

Federal Power Act

The Federal Power Act of 1935 extended the reach of the Federal Power Commission to cover the electricity industry. Specifically, the act gave the Federal Power Commission, and later FERC, authority to license companies to generate and transmit power in interstate service and to ensure that the terms and conditions are reasonable and nondiscriminatory. This act instituted the division between state and federal authority over electricity markets.

Public Utility Holding Company Act

Congress enacted the Public Utility Holding Company Act (PUHCA) in 1935 to transform large, complex holding companies into simpler structures. Before PUHCA was passed, three holding companies controlled over half of the generation in the United States. PUHCA gave the Securities and Exchange Commission the broad authority to break up the large holding companies by requiring them to divest holdings; become consolidated within defined geographic areas as single, integrated utilities; and have intra-holding-company agreements subject to oversight. In large measure, the purpose of PUHCA was to prevent utilities from using intracorporate cost shifting and other institutional abuses that thwarted the ability of state regulators to set reasonable prices. This transformation did not occur easily, because it took about 15 years to settle legal challenges to the act. But PUHCA essentially created the typical structure in which a utility provided generation, transmission, and distribution within a single state.

Public Utilities Regulatory Policies Act

More than 40 years ensued before the next major piece of federal electricity legislation. Spurred by the energy crises of the 1970s, Jimmy Carter's administration enacted a package of five bills under the title of the National Energy Act of 1978. The most significant part of the act for the electricity industry was the Public Utilities Regulatory Policies Act (PURPA). PURPA established a class of nonutility generators called *qualifying facilities* (QFs), which includes small generators using non-fossil-fuel *renewable* energy sources, and *cogenerators*, which can supply electricity generated as a by-product of industrial production processes. It required utilities to connect QFs to the transmission grids and purchase QF power at a price not exceeding the "avoided costs" of not having to produce that power themselves.

States were left to determine how great the avoided cost would be. Some states aggressively applied PURPA by requiring utilities to enter into contracts 20 or 30 years in length with prices that turned out to be extremely high. The indirect consequences of PURPA were to create an independent generation sector and lay the foundation for future competitive generation markets.

Energy Policy Act

Fourteen years after Congress opened up the wholesale market to QFs, in 1992 it passed the Energy Policy Act (EPAct), which set in motion the process of allowing power producers to compete to sell their power to local distribution utilities. First, it established a new category of nonutility generators known as *exempt wholesale generators* (EWGs), which are exempt from the requirements of PUHCA. EWGs—which can be owned by a utility, a holding company, or a company not affiliated with a utility—are not required to meet PURPA's renewable fuel and cogeneration requirements, and utilities are not required to purchase power generated by EWGs.

Second, EPAct amended the Federal Power Act of 1935 by giving FERC the authority to order utilities to provide transmission service, or wheeling, of wholesale power, on a reasonable and nondiscriminatory basis. Both buyers and sellers could now petition FERC to direct transmission-owning utilities to provide wheeling service at just and reasonable rates to recover transmission costs. FERC's Orders 888 and 889, issued in the spring of 1996, implemented these mandates (see box). EPAct preserved the Federal Power Act's jurisdictional separations by prohibiting FERC from ordering retail wheeling (i.e., providing transmission access directly to end users); these decisions remain in the hands of state legislatures and PUCs.

FERC Orders

FERC—empowered by the Congress under EPAct—issued Order 888 in April 1996 to require that utilities open their transmission grids for access by all generators on nondiscriminatory terms and conditions. If transmission capacity is insufficient to handle a request for transmission service, utilities are required to expand transmission capacity to accommodate that request. Order 888 also required "functional unbundling" of wholesale power prices, requiring utilities to separate the rates for generation, transmission, and ancillary services. It also included guidelines for the implementation of *independent system operators*, separate entities that would manage utility-owned transmission lines in such a way as to prevent discrimination against unaffiliated power producers in the ability to wheel power. Supplementary Order 889 required that information on transmission capacity, ancillary services, and prices of transmission be provided on a common information network, the Open Access Same-time Information System.

In response to claims that Order 888 did not go far enough to ensure open access, FERC issued Order 2000 in December 1999. This order requires transmission-owning utilities regulated by FERC to make certain filings regarding participation in *regional transmission organizations* (RTOs). The order provides minimum characteristics and functions that a transmission entity would need to satisfy to become an RTO. Although participation in RTOs is technically voluntary, FERC considers the formation of RTOs as a means to create competitive wholesale generation markets by removing engineering and economic inefficiencies and by reducing opportunities for transmission owners to discriminate. (See Chapter 7 for more on independent system operators and the discrimination issue.)

Price Variations

During the past decade, EPAct, FERC's Orders 888 and 889, and subsequent actions in some states have allowed competition to replace regulation as the primary method of determining electricity prices. This development is the result of a confluence of forces, some of which have been unfolding during an even longer period. At the top of the list is the variation in prices for electricity across the country. Price variations suggest that those paying high prices could benefit from greater competition in the provision of power.

Power rates differ in many ways, and for many reasons. Rates differ between industrial and residential customers, among states, and from one power provider to another.

Industrial versus Residential

Industrial customers typically pay lower prices than other classes of customers (see Figure 3-1). Partly, this is because industrial users have a less variable demand level and thus contribute less to the need for high-priced peaking power. Some industrial

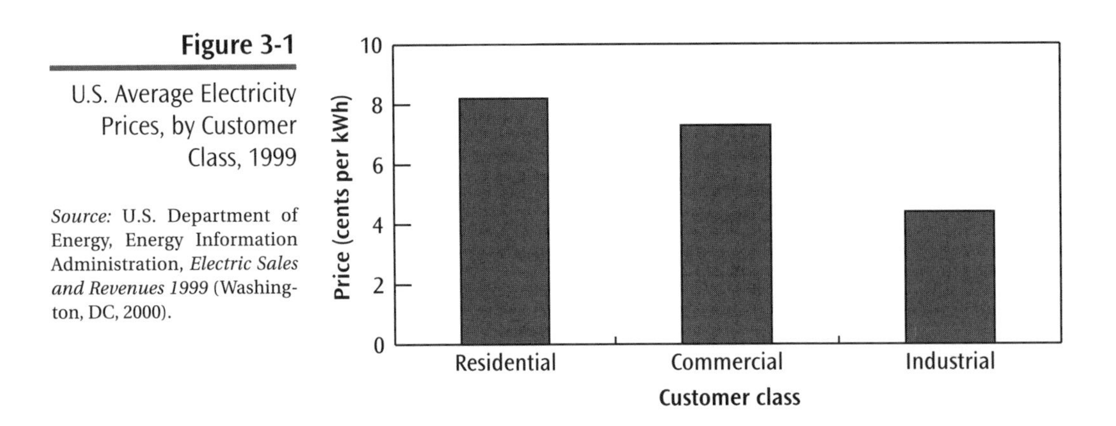

Figure 3-1

U.S. Average Electricity Prices, by Customer Class, 1999

Source: U.S. Department of Energy, Energy Information Administration, *Electric Sales and Revenues 1999* (Washington, DC, 2000).

customers can receive electricity at high voltages, hence reducing the cost of delivery. Industrial customers also have more options in purchasing electricity from the local utility, including self-generation and even plant relocation, which allow them to extract lower prices from utilities and regulators. Allowing retail competition for all classes of customers could provide more alternatives for residential and commercial customers, thereby helping to reduce their electricity prices.

States versus States

Consumers in some states, particularly New York, the New England states, and California, pay as much as double the national average price for power (see Figure 3-2). It is not surprising that these regions have been among the first to open markets to retail competition, in the belief that their consumers could benefit by getting less expensive power from other states. In general, prices are lowest in the Northwestern states and highest in the Northeastern states, Alaska, and Hawaii.

Differences across states are attributable to differences in resource base, in the mix of utilities serving a region, and in regulatory programs. Many of the Northeastern states face high prices because of costly long-term power purchase contracts with qualifying facilities under PURPA and costly regulatory programs to promote conservation and renewables.

Differences across Power Providers

Under the cost-based regulatory system by which IOUs have been regulated, differences in cost will be reflected in electricity prices. These nominal cost differences may result only from historical factors; for example, one utility's older plants may have been depreciated more than another's newer plants. Under competition, arbitrary historical differences typically do not affect prices; all sellers get the same price.

In addition, IOUs typically charge on average slightly higher rates than publicly owned utilities or cooperatives. Tax policies and power preferences underlie some of the differences in price among the different types of utilities. Publicly owned utilities and cooperatives can issue tax-exempt bonds to finance their investments and therefore enjoy lower costs of capital than investor-owned utilities. Publicly owned

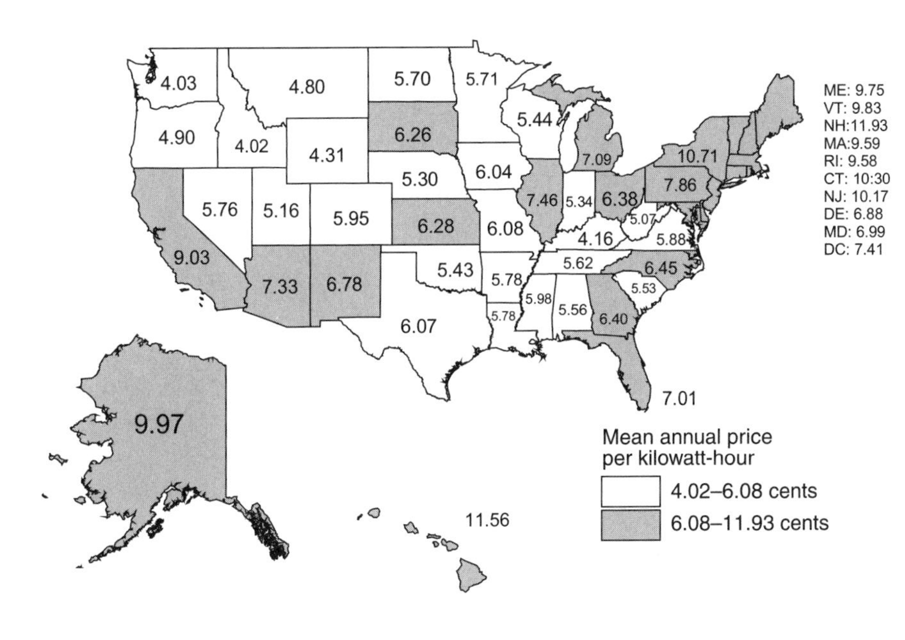

Figure 3-2

Annual Electric Utility Average Revenue per Kilowatt-Hour for All Sectors by State, 1990–2000

Source: U.S. Department of Energy, Energy Information Administration. http://www.eia.doe.gov/cneaf/electricity/page/sales/annual_ave_prices.xls (accessed January 20, 2002).

utilities and cooperatives also have priority access to low cost federal hydroelectric power, which further contributes to their lower electricity prices. (For more on the role of public power in a restructured electricity industry, see Chapter 13.)

Other Factors Encouraging Competition

Other major drivers for the move to greater competition have been:

- *Technological change.* Advances in generation allow small-scale plants to match the cost advantages of larger plants. This makes it easier for firms to raise capital and get enough business to compete effectively against established utilities.
- *PURPA and feasibility.* Although PURPA has been criticized for forcing utilities into long-term contracts to purchase power at very high prices, a fortunate by-product was that it demonstrated that independent generators could supply power to an otherwise vertically integrated, closed transmission and distribution system without causing significant disruption.
- *Green power.* Policymakers, environmental activists, and power entrepreneurs all view retail competition as creating opportunities to sell power at a premium to consumers who are willing to pay more to pollute less. Whether or not electricity restructuring increases or decreases pollution depends on numerous factors, primarily where and with what fuels electricity would be produced in an open market (see Chapter 15).
- *Successful precedents in other industries.* Deregulation has probably been the leading industrial policy in the United States in the past 25 years. The efforts were most successful in industries such as banking, trucking, and airlines, where competition had been most clearly suppressed. AT&T's divestiture of its local telephone monopolies led the move to transform long-distance telecommunica-

tions, including the Internet, into a competitive business. Increasing numbers of state energy regulators are allowing customers to choose their natural gas provider. Electric power is one of the last major U.S. industries to remain largely regulated, and proponents of expanding competition believe that similar benefits could be gained in this $200 billion sector.

Federal Legislative Proposals

Although FERC is responsible for promoting wholesale competition, most of the legislative and regulatory activity to promote retail competition is happening in the states. The next two chapters detail some of those efforts. However, several federal lawmakers and industry participants have expressed concerns about the lack of uniformity that results from state-by-state decisions. They are also concerned about existing federal policies that pose potential barriers to establishing a level playing field among all competitors.

Before the developments in California in late 2000, about two dozen federal bills related to electricity restructuring had been introduced in either the House or Senate. These bills varied in focus and scope, but they shared many common elements. Many of the bills included provisions that would require states to establish retail competition by a particular date, whereas at least one took a more cautious approach of allowing states to opt out of that requirement if it could be shown that the current system or an alternative policy would be more beneficial for consumers within the states.

The bills also addressed several of the issues delineated in later chapters of this book. Concerns regarding system reliability, discussed in Chapter 11, would be addressed through the creation of an electricity reliability organization with FERC oversight. A common thread in these proposals was to make all utilities "exempt wholesale generators" by repealing PUHCA's restrictions on interstate operations and PURPA requirements to purchase power from cogenerators and renewable fuel users. Some bills also gave FERC jurisdiction over the merger or consolidation of holding companies and generation-only companies, provided FERC with authority to remedy market power (see Chapter 9), and extended FERC's jurisdiction to operations of federal power administrations, municipal utilities, and cooperatives, similar to regulatory oversight of IOUs. Most of the proposals also included some type of provision to encourage greater use of renewables (see Chapter 16) and guaranteed the recovery of stranded costs, losses that would arise if retail electricity prices generate insufficient revenues to compensate investors for the costs of building generators or purchasing power under PURPA-mandated contracts (see Chapter 14).

Despite these calls for federal legislation, the major drivers for competition are pushing states forward to implement forms of retail competition while the national reliability concerns are left to federal policymakers. In the next two chapters, we examine the directions taken by California and Pennsylvania. Because the United States is not the first country to take on the restructuring issue, we also provide brief summaries of experiences in Chile and the United Kingdom.

CHAPTER

International and U.S. Restructuring Experiences

Although the United States is a leader in efforts to introduce competition into several formerly regulated industries—including telecommunications, banking, and transportation, particularly at the federal level—it has not been an early mover in opening retail electricity markets to competition. Numerous other countries, including Argentina, Australia, Bolivia, Canada (Alberta), Chile, Colombia, New Zealand, Norway, Peru, and the United Kingdom, have initiated restructuring toward more open markets, sometimes including customer choice of electricity suppliers. In the United States, all of the activity has been at the state level. As of September 2001, 23 states had enacted comprehensive legislation to set retail electricity competition in motion, and one additional state plus the District of Columbia had issued comprehensive regulatory orders toward the same end. Although, in light of its recent electricity market crisis, California has ended retail competition and a few other states are in the process of reconsidering it, most of the other states that have decided to adopt competition are sticking by that decision, at least for the time being.

All of these experiences offer lessons to teach others just setting out on the road to electricity competition, or deciding whether or not to begin the journey. We briefly examine electricity restructuring in Chile and in England and Wales. Then we discuss the initial models for competition adopted in two of the states that were early adopters of competitive retail electricity markets: Pennsylvania, where electricity markets have worked fairly well, and California, where they have not.

Models from Other Countries

The United States is not the pioneer among nations when it comes to electricity restructuring efforts. Recently implemented transmission models in the United States, such as the PJM (Pennsylvania–New Jersey–Maryland) and California independent system operators (ISOs), are modifications of models implemented in other countries. Although state policymakers are closely observing the markets in California and Pennsylvania, unresolved issues relating to the performance of foreign electricity models are of interest to domestic policymakers as well.

Chile

Chile is considered to be the pioneer in global electricity restructuring. The precursor of this electricity restructuring effort was a broad initiative in the 1970s to privatize Chilean enterprises, farms, and banks. Subsequently, a series of legislative rules in the early 1980s, primarily the Chilean Electricity Law of 1982, *privatized*—transferred to private owners—the government-owned utilities (see box). These laws eliminated monopoly franchises and allowed large customers to purchase wholesale electricity from any generator or distribution company.

The Chilean electricity industry now comprises two main interconnected systems, the Sistema Interconectado Central (SIC), which supplies more than 90% of the power (mostly hydroelectric power) consumed by Chilean residents, and the Sistema Interconectado Norte Grande, which serves mostly mining and industrial demand with power generated by fossil fuels. Generators yield control to the Economic Load Dispatch Center (CDEC) in each interconnection. The CDEC, an independent dispatch operator, plans and coordinates the operation of plants to attain least cost dispatch, under which plants with lower variable costs are dispatched before plants with higher variable costs of generation.

Generation companies sell electricity in three markets:

- *Unregulated market.* These include short- and long-term contracts, mostly to industrial and mining companies with a demand greater than 2 megawatts (MW). Customers negotiate directly with generating or distribution companies. "Retail competition" is only available for these large customers.
- *Regulated market.* This market is made up of consumers whose demand is 2 MW or less. Sales by generating companies to the distribution companies are regulated by the Ministry of the Economy, with the National Energy Commission providing regulated prices every six months based on forecasts of the projected power costs. Sales by distribution companies to final customers are regulated with tariffs varying by company and across customer classes.
- *Spot market.* Excess power generated by producers already meeting contractual arrangements specified above is transferred to generators not able to meet contractual arrangements. The CDEC arranges the spot price for this "balancing mechanism," which is based on the marginal costs of generation (see Chapter 9 for discussions of marginal costs).

Chile has implemented a form of yardstick regulation for distribution, in which the cap on one firm's prices is based on the costs of a hypothetical reference company designed to resemble other similarly situated distribution companies. Under yardstick regulation, if a distribution utility is more productive than the reference firm, it keeps the profits; if not, it loses. This form of regulation has resulted in lower distribution tariffs and is considered a positive aspect of Chile's restructuring. Transmission companies are required to provide open access at regulated rates to all generators. The privatization process did not include any electricity-specific laws addressing market power, because three companies own about 93% of the generation capacity in the SIC. In addition, the largest generator in the SIC also owns a majority of the transmission system, raising concerns about the efficacy of the open access rules (see Chapter 7).

Companies have questioned spot prices, dispatch models, and transmission procedures in the CDECs. Exacerbating the controversy, a severe drought in 1998–1999 disrupted the normal operations of the SIC due to its dependence on water for hydropower generation. A bylaw promulgated by the Ministry of Mining in 1998 increased the number of participants in the CDECs, while enabling the CDECs to become more autonomous, with decisionmaking power moved from participants to independent staffing groups advising the CDECs. The Chilean Congress passed a law in 1999 to further define reliability standards and compensation rules in the CDECs. For example, a generator cannot fail to meet its payment obligations by claiming that a drought situation is a force majeure event. Many of the changes implemented by the congressional law were motivated by the extreme drought conditions.

The lingering effects of the drought still affect the industry. The large fines facing generators when power is interrupted are discouraging new investment in generation projects. The power regulator in Chile is immediately trying to prepare legislation that would address the fines, power cuts, and rationing. With Chilean officials suggesting that the industry is in an electricity shortage situation, the government is considering building generation capacity. Finally, the regulator is in the early stages of planning a transition from least-cost dispatch in the CDECs to spot markets as the primary method for wholesale electricity exchanges.

United Kingdom

The restructuring of the United Kingdom's electricity industry began during the 1980s amid privatization in other sectors, including aerospace and telecommunications. In 1983, the process began with the passage of the Electricity Act of 1983. It gave independent power producers access to the trans-

Privatization

One difference between U.S. restructuring efforts and those of other countries is that the latter are frequently portrayed in terms of *privatization*, the transferring of publicly owned enterprises to private owners and investors. In the United States, most regulated industries with franchised monopolies are privately held and publicly regulated. The Postal Service, urban mass transit, and municipal electric cooperatives are the exceptions that prove the rule.

In most other countries, the situation is quite different. Electric utilities, along with telecommunications and other industries, are or have been owned by the state rather than private firms and investors. Consequently, privatization has preceded any type of regulatory reform or the introduction of competition in most economies throughout the world. In Chile and the United Kingdom, for example, privatization was an important and necessary step in electricity restructuring.

Privatization often seems to be a synonym for introducing competition, but the two processes are inherently quite different. Privatizing a natural monopoly may not ensure that entry will be attractive. In the United States, distribution and transmission will continue to be privately owned, but regulation rather than competition will be setting prices in those segments of the electricity industry for the foreseeable future. In general, the choice between the regulation of a privately owned monopoly and outright state ownership depends on fairly subtle factors. The most notable may be the relevance of nonmonetary goals (e.g., providing "universal service" to low-income or rural households) and the ability of a regulator to ensure that a privately owned, profit-maximizing enterprise promotes them.

Nor is public ownership inherently incompatible with competition. In the United States, the Postal Service faces extensive competition in the delivery of urgent mail (e.g., FedEx). In electricity, as both the California and Pennsylvania examples show, municipal utilities will be expected to open their networks to competitors, particularly if they want to compete for customers in other markets themselves.

When "privatization" does transform a publicly owned unnecessary monopoly into a genuinely competitive industry, the benefits are likely to be substantial. However, the benefits are likely to come more from the introduction of competition itself. The importance of privatization is mostly in assuring new competitors that the state will not continue to favor and subsidize the firm it owns.

mission grid. Like PURPA in the United States, the 1983 law required that the Central Electricity Generating Board (CEGB, the state-owned generation utility responsible for electricity generation, transmission, and distribution) purchase generation from independent power producers at avoided costs.

Restructuring accelerated with the Electricity Act of 1989, which included a comprehensive plan to break CEGB into three generation companies (Nuclear Electric, National Power, and PowerGen), a distribution network comprising 12 regional electric companies (RECs), and a grid company (National Grid Company) to be owned by the RECs. The government auctioned equity shares in the RECs in 1990 and, beginning in 1991, shares in National Power and PowerGen. The power producers were required to sell all electricity into the centrally operated England and Wales Power Pool. However, more than 90% of the electricity sold into the pool was covered by financial hedging contracts, known as "contracts for differences."

Contracts for differences involve an agreed-upon strike price per kilowatt-hour for a specific quantity of electricity. If the pool price falls below the strike price, the party purchasing power from the pool, usually a local distribution company or power retailer, will pay the difference to the generator. If the pool price rises above the strike price, the generator will pay the purchaser the difference. These contracts mitigate the risk that market participants would face if prices for all transactions were set in the spot market.

Retail competition was slowly phased in, with industrial customers having the first opportunities to choose an alternative supplier of electricity. Only 5,000 large customers with a maximum demand of more than 1 MW were allowed to choose their electricity supplier in 1990. Then, all those customers with a maximum demand greater than 100 kilowatts were able to choose their electricity supplier starting in April 1994. The phase-in process for all customers was finally completed in May 1999.

As of March 2001, roughly 25% of all residential customers had switched suppliers. About 56,000 residential customers a week are permanently switching away from their electricity supplier. Restructuring in the United Kingdom has resulted in a decrease in real electricity prices for residential, commercial, and industrial customers. For some customers, prices have fallen by 30% in real terms since the start of competition in 1990. Some critics suggest that of the three classes of customers, residential customers have received the lowest realized cost savings from restructuring.

Although restructuring may have given customers more choices and lower prices, the transition has been controversial. During the 1990s, the U.K. regulator—then the Office of Electricity Regulation, and now the Office of Gas and Electricity Markets (OFGEM)—issued several reports regarding dominance of National Power and Powergen in the England and Wales Power Pool. OFGEM provided evidence of possible "gaming" of pool rules by National Power and PowerGen; in addition, the generators dominated the setting of the pool price.

After noticing a sudden increase in the pool prices in 1992, OFGEM negotiated with National Power and Powergen to sell nearly 15% of their generation capacity in 1993. The rationale for these divestitures was to improve competition by expanding the number of independent suppliers. In addition, the RECs were required to divest the National Grid in 1995. The government continued to hold Nuclear Electric until 1996, when some of the relatively modern nuclear plants were privatized. The older plants are still owned by British Nuclear Fuels Limited, which is ultimately owned by the government. The plants account for about 7% of U.K. generation.

In late 1997, the U.K. government expressed concern about the way wholesale electricity prices in the pool appeared to be manipulated by large generators, particularly during peak hours. In response to this concern, the regulator conducted a review of electricity trading arrangements in the pool. As a result of that review, in March 2001 OFGEM launched new rules, known as the New Electricity Trading Arrangements, to replace the traditional pool structure. A primary goal of these new rules is to allow for greater trading of electricity outside of a central power market. The new arrangements affect electricity transactions among generators, suppliers (as electricity retailers are know in the United Kingdom), and electric power brokers.

The exchanges of power will center on three types of markets: a forward and futures market, a short-term bilateral market, and a real-time balancing mechanism. The first two markets will be largely decentralized, allowing traders to freely negotiate electricity deals. The expectation is that 90% of all power transactions will take place in the forward market. The balancing mechanism, operated by the National Grid Company, is needed to price and settle the surpluses and deficits that result from mismatches between power contracted to be sold and the amount actually generated due to the physical exchange characteristics of electricity. An independent company, Elexon, has replaced the pool by managing the Balancing and Settlement Code that establishes the rules for exchanges and governance. A recent report by OFGEM indicates that electricity prices fell during the first three months of operation under the new rules. As was expected before the launch, the bulk of electricity has been traded in the forward and futures market.

Transmission and distribution remain controlled by the government under price cap regulation, by which rates are adjusted upward automatically for inflation, but with an annual percentage reduction put in place on the basis of expected productivity. As we discuss in Chapter 8, price cap regulation encourages the regulated firm to cut expenses, while reducing the need for regulators to monitor the wire companies' actual expenses.

OFGEM is continuing to modify its regulation of the National Grid Company and distribution companies. Under the new market arrangements, OFGEM regulates rates to recover the costs of transmission investment and maintenance separately from the costs incurred for balancing power exchanges between buyers and sellers. Contracting for reserve power and forecasting load are some of the costs associated with balancing power. Price cap regulation remains in effect for the transmission asset side. For National Grid Company's system-operation functions, the regulator has established a set of cost targets. If the company can reduce costs below the target, it keeps some of the cost savings and returns the rest. When its costs are above the target, National Grid pays the difference.

Distribution companies are currently under price cap regulation, but OFGEM is considering new ways to strengthen incentives to provide high-quality service. One way this will be achieved is by publicly reporting service quality measures for different distribution companies. Under this approach, customers would be able to easily make comparisons among distribution companies on various dimensions of service quality.

U.S. Models

In the United States, the introduction of competition has taken different paths in different states (Figure 4-1). In this section, we discuss two models that characterized the initial policies to restructure electricity markets in California and Pennsylvania. In the next chapter, we will discuss more recent events in California that culminated in the decision to suspend competition there.

California

Among the states, California became one of the pioneering national leaders in the move toward retail competition. The start began in an environment with high electricity rates and concerns about the future of the state's economy given recent losses in manufacturing jobs from defense cutbacks. At the time of the initial attempt to investigate electricity restructuring in 1992, the average price of electricity was 6.8 and 9.7 cents per kilowatt-hour (kWh) in the United States and California, respectively. Major reasons for the high electricity rates of investor-owned utilities (IOUs) in California included both the investment in costly nuclear generation capacity and the large amounts of generation from qualified facilities, which utilities were required to purchase under high-priced, long-term contracts.

Paving the Way for State-Level Restructuring. In 1992, the California Public Utility Commission (CPUC) asked its staff to prepare a report that would examine the current electricity industry and future trends and provide alternative regulatory approaches that could be adapted to the evolving industry. This report, known as the "Yellow Book" and published in early 1993, concluded that the current regulatory structure was not suited to govern the state's electricity industry and provided four general reform strategies, one of which included a model of limited customer choice. Panel hearings were held in the spring and summer of 1993 to explore additional perspectives on possible regulatory reform.

On the basis of the Yellow Book, panel hearings, and comments received from various stakeholders, in the spring of 1994 the CPUC issued a proposal of regulatory reforms for California's electricity industry, known as the "Blue Book." The Blue Book proposed to generally open the California electricity industry through *functional unbundling*, which intended to separate a utility's regulated transmission and distribution services from its competitive generation and marketing businesses. The CPUC also proposed to replace cost-of-service regulation with incentive-based regulation (described in Chapter 7; also see Chapter 9 on cost-of-service regulation). After the CPUC conducted extensive debates and numerous stakeholder meetings to address the transition from closed and regulated markets to open and competitive ones, it voted to adopt a final restructuring proposal in December 1995. But implementation of the initiatives put forth by the CPUC required some changes to existing state laws.

In 1996, the California legislature passed and Governor Pete Wilson signed into law a comprehensive electricity restructuring bill (AB 1890) that generally followed the Blue Book's recommendations. Under AB 1890, retail competition was scheduled to begin in January 1998. AB 1890 included a five-year plan that would give residential and small commercial customers a 10% rate reduction below 1996 prices, which would then be frozen from January 1998 through December 2002, or until a

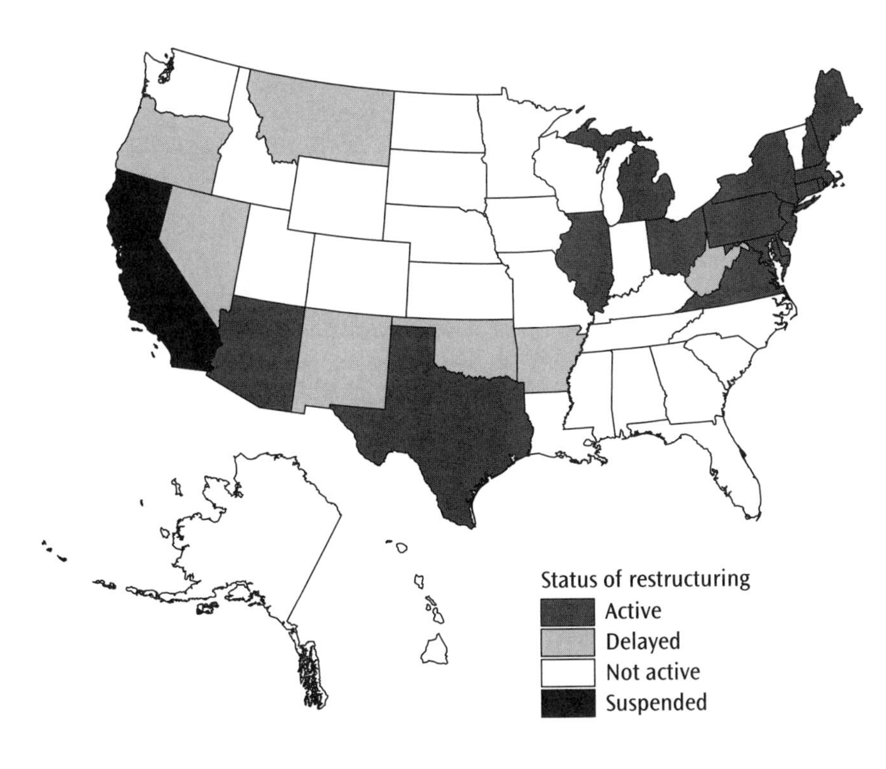

Figure 4-1

Status of Electricity Restructuring, January 2002

Source: U.S. Department of Energy, Energy Information Administration, http://www.eia.doe.gov/cneaf/electricity/chg_str/regmap.html (accessed January 8, 2002)

utility had recovered all of its stranded costs (see below and Chapters 1 and 14 for discussions of stranded costs).

Following the passage of AB 1890, the CPUC made a number of major decisions in implementing the legislature's mandate:

- *Creation of a power exchange and independent system operator.* The CPUC decision set up an independent, centralized energy auction matching bids for and sales of power, now known as the California Power Exchange (PX). Also, the CPUC created the California Independent System Operator (ISO) to operate the transmission networks owned by the state's three major utilities. The ISO is responsible for load balancing and reliability (see Chapters 10 and 11). It also was charged with operating markets for the provision of ancillary services (Chapter 10) and managing transmission congestion (Chapter 8). During the five-year rate freeze implemented by AB 1890 or until they recovered their stranded costs, IOUs were required to purchase the power they need to supply customers who chose not to switch to a competitive supplier through the PX. The IOUs were also required to sell into the PX market any electricity they produced or purchased through continuing "qualifying facility" contracts (discussed with PURPA in Chapter 3).
- *Recovery of stranded costs.* The CPUC provided for the recovery of *stranded costs*—unrecovered capital costs of generators and long-term contracts to purchase power under PURPA (Chapters 3 and 14)—through a competitive transition charge (CTC). The transmission and distribution costs (along with the average market price of generation as determined by the PX) are subtracted from the frozen, bundled rates to arrive at the CTC for the IOU. As a result, the CTC goes

Securitization

In electricity restructuring, securitization refers to the use of state-issued bonds to raise the funds to pay utilities for their stranded costs. As a result, utilities do not have to rely on a continuing regulatory bargain to collect their stranded costs. Instead, the utilities are paid up front for their stranded costs, and the state is responsible for paying the interest and principal on the bonds from the revenues collected through the CTC or other fees earmarked for stranded cost recovery.

A numerical example may help explain both how securitization works and why a state might adopt it. Suppose a utility had spent $1,000 on a generator while the industry was regulated and the utility expected to be able to recover its investment through the revenue it would make from electricity sales. However, after competition, the electricity price would be too low to generate these revenues, and the utility would get back only $700, leaving $300 in stranded costs.

Suppose that the state decides that the utility should recover this $300. (For more on the policy issues associated with this decision, see Chapter 14.) Normally, this would be done by letting the firm collect the money over time, including a market rate of interest. Suppose that doing so would result in an annual payment—principal on the $300, plus interest—of $100 a year for four years, and that this would require a CTC of 4 cents per kWh.

But there is an alternative. The state could "securitize" the stranded costs by issuing bonds to raise the $300 now, and give that money to the utilities. Revenues from the CTC would be used to pay off the principal and interest on these state-issued bonds. If the interest rate on those bonds were the same as the market interest rate used to calculate the CTC above, the annual payments and CTC would be identical.

However, the interest on state bonds, unlike interest payments generally in the market, is exempt from federal income tax. Therefore, the interest rate on bonds will be below the market rate. Instead of $100, the annual payment might be only $90, with a CTC of 3.6 cents per kWh instead of 4 cents. In effect, the costs to the state and its ratepayers fall not because of a real economy, but because the federal treasury absorbs the costs in reduced income tax payments. However, the reduced costs to the state from securitization contributed to enabling California to institute its initial 10% reduction in retail rates.

up when PX prices go down, and vice versa, leaving the IOU's final electricity sales price constant.

- *Securitization.* AB 1890 added to the CPUC proposal by allowing bonds to be issued in order to raise the funds to pay utilities for stranded costs through a process called *securitization* (see box).
- *Divestiture of generation assets.* To reduce concerns about market power in wholesale generation, the CPUC directed the two largest IOUs, Pacific Gas and Electric (PG&E) and Southern California Edison (SCE), to divest 50% of their fossil-fuel generation capacity. PG&E, SCE, and San Diego Gas and Electric (SDG&E) have since divested all fossil-fueled generators.
- *Unbundling of prices.* Prices charged by the IOUs are unbundled to separate generation charges from transmission and distribution charges. The CPUC included plans to implement incentive regulation for regulated distribution tariffs.
- *Public power participation.* Public power systems (described in Chapters 3 and 13) were given the choice of whether or not to allow retail choice in their service territories. Competitive transition charges cannot be collected by public power utilities unless they yield control of their transmission facilities to the ISO.
- *Funding public benefits.* The legislation supported the CPUC's proposal to finance energy efficiency, low-income renewable energy, and research and development technology programs. The benefits would be paid by a "nonbypassable" charge paid by retail consumers (see below and Chapter 14 for discussions of nonbypassable charges).

Following a three-month delay due to technical problems with the ISO and PX, the CPUC began retail competition (known as Direct Access in California) in March 1998. Direct Access allows all customers—industrial, commercial, and residential—to purchase power from an electricity service provider (ESP)

Table 4-1

Customers Switching to Electricity Service Providers by Class in California under Direct Access, as of July 31, 2000

Class of Provider	*Percentage of Customers*	*Percentage Load*
Residential	1.7	2.8
Commercial (less than 20 kilowatts)	2.8	4.3
Commercial (20–500 kilowatts)	6.0	13.2
Industrial (more than 500 kilowatts)	13.5	28.1

Source: Direct Access Reports, California Public Utilities Commission.

instead of from the IOU. The ESPs handle the financial aspects of purchasing power from the generation market and selling to final customers while relying on the distribution wires of the utility distribution company, the incumbent IOU, to physically deliver the power. ESPs, or aggregators that act as intermediaries between customers and ESPs, may combine the demands from a number of customers to obtain lower rates and special features. Customers who do not select an ESP may continue to purchase power from the local utility distribution company (UDC), which basically acts as a reseller of power purchased from the PX.

How Did Consumers React to Competition? Just as restructuring was scheduled to begin, opponents attempted to reverse its course through the ballot box. In the summer of 1998, a coalition of consumer groups that favored lower prices and opposed stranded cost recovery and nuclear power challenged AB 1890 by collecting more than 700,000 signatures to add Proposition 9 to the November ballot. Proposition 9, among other things, would have imposed a 20% residential rate decrease and prohibited stranded cost recovery and securitization for nuclear generation. After a hotly contested campaign, AB 1890 survived and Proposition 9 failed, as 73% of the electorate voted against the initiative.

As the competitive market got started, differences existed in how various customer classes engaged retail choice. By the end of 1998, 78,881 residential customers, less than 1% of those served by UDCs, switched to an ESP. By July 2000, the figure had risen to only 152,023 residential customers, or 1.7% of the UDC business. However, the percentage of customers switching suppliers was greater in other classes. For example, Table 4-1 shows 13.5% of all industrial customers served by UDCs had switched to an ESP as of July 2000. The corresponding amount of power purchased represented 28.1% of the total power purchased by industrial customers served by UDCs.

The initially mandated 10% rate decrease for UDC residential customers provides a partial explanation of the low numbers of customers switching to an ESP. Also, as a result of the mandated price cap, customers had limited opportunities to save money by switching suppliers. In addition, many households may have found it too costly to wade through a complicated and detailed menu of pricing options, particularly when some final pricing outcomes were not exactly known due to a variety of factors.

Pennsylvania

The introduction of retail electricity competition in Pennsylvania began with a Pennsylvania Public Utility Commission (PUC) symposium on retail competition in November 1993. About three years later, Governor Tom Ridge signed into law the

Table 4-2

Major Differences in State-Level Electricity Restructuring Policies

Issue	*California*	*Pennsylvania*
Implementation of retail choice	Full-scale implementation	Phased in with pilot program
Market organization	State creation of separate power exchange and independent system operator (ISO)	Nonmandated creation of integrated power exchange and ISO from existing power pool
Rate caps	10% mandated statewide rate decrease during transition period	Generally smaller mandated rate decrease during transition period; varies by utility
Stranded costs: timing	Shorter collection period	Longer collection period
Stranded costs: recovery method	Variable "competitive transition charge"; varies inversely with wholesale power price	Fixed surcharge added to transmission and distribution wire charges
Incumbent's retail power price	Determined by wholesale market price as long as below retail rate cap less wire charges	"Price to compare" equals rate cap minus wire charges and stranded cost contribution

state's Electricity Generation Customer Choice and Competition Act. Some of the provisions of this Pennsylvania law—such as allowing municipal utilities to opt out of retail choice, allowing a *nonbypassable charge* to fund universal service and energy conservation (or public benefits), and allowing for the PUC to implement incentive regulation for distribution—are also found in the California law. The major differences between the Pennsylvania restructuring plan and the California model include how policymakers addressed implementation of retail choice, recovery of stranded costs, and organization of generation and transmission functions (see Table 4-2).

From Pilot Programs to Implementation. The first step in implementing retail competition in Pennsylvania was a requirement that its IOUs submit proposals for pilot programs under numerous guidelines. The major guidelines included the following:

- The size of each IOU's pilot program should be roughly equivalent to 5% of the IOU's peak load for each customer class, and the pilot program is to last for at least one year.
- Technical and operational guidelines for electricity suppliers were to be specified.
- Utility tariffs shall contain an unbundling of generation from jurisdictional transmission and distribution.
- Utilities should offer distribution service and any surplus generation to affiliates on the same terms and at the same prices that they offer these services to non-affiliates.

- Methods for recovering stranded costs could be included.
- Utilities were required to address how they would promote customer education, safety, and reliability.

The open enrollment period for customer participation in the pilot programs resulted in more willing participants than the number needed for the program. Lotteries were held to randomly select participants. The first pilot program began in November 1997, and by the end of the spring of 1998, about 230,000 customers had retail choice, the opportunity to select an alternative electricity supplier. The pilot program eased utilities, consumers, and new ESPs into retail competition. The phase-in process for comprehensive retail access was subsequently accelerated, with all customers of the IOUs getting retail choice before the initial January 2001 deadline for full implementation.

As was the case in California, customers receive the power through the distribution lines of a UDC, but can purchase electricity from the existing supplier or select an ESP. The promotion of Electric Choice, the name for retail competition in Pennsylvania, has included mailing letters to customers that explain how to shop for electricity from ESPs. But in contrast to California, more customers in Pennsylvania have selected ESPs. One major factor may be an easier method for determining if switching to an ESP saves money.

In Pennsylvania, customers shop for electricity by comparing the "price to compare" of the UDC (also known as a "shopping credit") to the price to compare of the ESP. For the UDC, the price to compare is set close to the incumbent utility's embedded cost of generation (i.e., the average cost of electricity, taking into account both fixed and variable costs). For example, if a customer currently faces a UDC price to compare of 5 cents per kWh, the customer would shop for a price to compare lower than 5 cents per kWh from an ESP. If the price to compare from the ESP is 4.5 cents per kWh and the customer uses an average of 750 kWh a month, the savings would be $3.75 a month. Additional monthly fees or special services may be applicable, but this shopping program defines the Pennsylvania approach to retail choice.

Pennsylvania's approach to retail choice initially provided incentives for customers to switch to ESPs, as is indicated by Table 4-3. For example, nearly 16% of

Table 4-3

Percentage of Utility Distribution Company Customers Switching to Electricity Service Providers by Class in Pennsylvania, as of July 1, 2000

Utility Distribution Company	*Residential*	*Commercial*	*Industrial*
Allegheny Power	0.6	3.1	8.5
Duquesne Light	29.4	5.5	3.0
GPU Energy	4.1	9.7	15.8
PECO Energy	15.8	30.2	44.3
Penn Power	6.4	2.4	8.8
PP&L	2.3	12.3	8.7
UGI	3.4	1.3	None

Note: Pennsylvania also has three other distribution-only investor-owned utilities, which collectively provide retail choice to 17,000 customers. Data on customers who switched to electricity service providers from these utilities were not available.

Source: Pennsylvania Office of Consumer Advocate.

1.3 million residential consumers served by PECO Energy's distribution lines had switched to an ESP as of July 2000, even though PECO implemented an 8% monthly rate reduction for all consumers in 1999 (which later was reduced to 7%). However, the willingness of customers to sign on with ESPs has waned a bit recently. Between April and June of 2001, the total number of Pennsylvania customers purchasing electricity from ESPs declined by nearly a third, as some ESPs went out of business.

Stranded Costs and the "Price to Compare." As in California, the stranded cost burden to Pennsylvanians is reduced through securitization, but Pennsylvania regulators have typically given utilities a longer time period to collect stranded costs. However, there are important differences between the two states in how the stranded cost surcharge or CTC is calculated.

In Pennsylvania, regulators determined transmission and distribution costs along with the charge for stranded cost recovery and for universal service and energy conservation. The nonbypassable universal service and energy conservation charge is embedded in the distribution rates for utilities. The CTC is determined by the total value of stranded costs, whereas the per-kilowatt-hour charge stays the same or decreases during the collection period as determined in the individual restructuring agreements. The unbundled price for the IOU's generation, or its price to compare, is determined by subtracting the sum of the CTC and transmission and distribution costs from IOU's frozen, bundled price for delivered power. By contrast, in California, the stranded cost portion of the CTC charge is the result of subtracting transmission, distribution, and power exchange prices from the mandated rate—freezing consumer prices as long as stranded costs were being recovered.

Because the prices to compare in Pennsylvania typically exceeded competitive wholesale generation prices, ESPs initially found it relatively easy to enter the market. In California, where wholesale electricity prices were determined in one centralized exchange until early 2001, it was difficult for ESPs to be able to offer electricity for less than the PX-determined prices. In Pennsylvania, if a substantial number of customers remain with the utility, it can recover stranded costs beyond what it is guaranteed through the settlement agreements, because the price to compare exceeds the competitive price of generation. By contrast, in California the ability of the CTC to fluctuate with the price of electricity means that IOUs cover stranded costs, either through the CTC when power prices are low, or through power price itself, when it is relatively high, as it has been for the last year or so.

No State-Mandated Power Exchange or ISO. Unlike in California, the Pennsylvania law did not provide for a centralized power exchange and an organization to manage transmission. A large power pool—responsible for coordinating electricity exchanges and maintaining reliability—already existed in the region. This power pool, formerly known as the PJM Power Pool, evolved into the PJM Interconnection, an ISO charged with operating the transmission grid and a centralized power market for electricity in an area encompassing much of Delaware, the District of Columbia, Maryland, New Jersey, and Pennsylvania. The PJM Interconnection facilitates access to the transmission grid for electricity producers wishing to sell electricity in the region, and it also operates a fairly liquid wholesale market in generation. Some Pennsylvania IOUs, mainly those located in the western part of the state, currently do not participate in the PJM Interconnection; however, a proposal to have PJM serve as the regional transmission organization for Allegheny Power, the holding company for three major western Pennsylvania utilities, has been approved by the FERC.

What Have We Learned?

The experiences of California and Pennsylvania, and of Chile and the United Kingdom internationally, show that restructuring national or local electricity industries is not easy, even when it works fairly well. From these experiences, we can glean some insight into the complexities of moving toward competition, but no general prescriptions that apply in all cases.

Foremost among these experiences is the imposition of higher rather than lower prices, accompanied by rolling blackouts and utility bankruptcies, which arose in California in the summer of 2000. In the next chapter, we will review the history of these important developments, identifying a list of 10 possible culprits related to supply conditions, market design, and market power. Our assessments reflect the reality that the factual issues, financial calamities, political controversies, and regulatory responses remain open, and all may arise in different forms in other states.

CHAPTER

The California Experience

The Crisis in a Nutshell

A first step in understanding the California electricity experience is to know that it worked reasonably well for more than two years. The concerns of regulators and researchers ran the gamut, including the procurement of ancillary services, possible noncompetitive pricing, transmission rate structure and congestion management, and dispatch policies of the independent system operator (ISO). In addition, the competitive transition charge (see Chapter 4) and retail price controls led to less entry than expected. However, from the standpoints of prices and reliability, restructuring in California was, initially, reasonably successful. Figure 5-1 indicates that until June 2000, electricity prices in California remained fairly low. Wholesale prices ranged roughly between 1 and 3.5 cents per kilowatt-hour off peak, and peak prices were roughly a penny higher.

A similar sense of the performance of the market can be gleaned from looking at the number of occasions in which reserves were declared to be precariously low, referred to by the California ISO as *staged emergencies*. Figure 5-2 shows that before the summer of 2000, such emergencies were virtually nonexistent; they occurred only during summer months, and three occurred, at most, in any one month. At no time during that period did blackouts related to systemic imbalances occur.

The character of the "crisis" changed over time. As the figures show, the turn for the worse began around June 2000. Peak and off-peak wholesale prices began to spike up to levels nearly 10 times those reached during the previous two years. Because San Diego Gas and Electric's (SDG&E's) retail rates had been deregulated, these wholesale prices led to reports of up to a tripling of retail electric bills relative to the summer of 1999, leading to the re-regulation of SDG&E's rates in September 2000, retroactive to the previous summer. Then came a threat of blackouts throughout the state (see Chapter 4). Although wholesale electricity prices ballooned, retail price ceilings on the IOUs left the utilities unable to cover their expenses. As summer passed into fall, the crisis became financial, as the accumulated difference between wholesale costs and retail prices led to a deficit of more than $14 billion that no one—utility stockholders, electricity consumers, taxpayers,

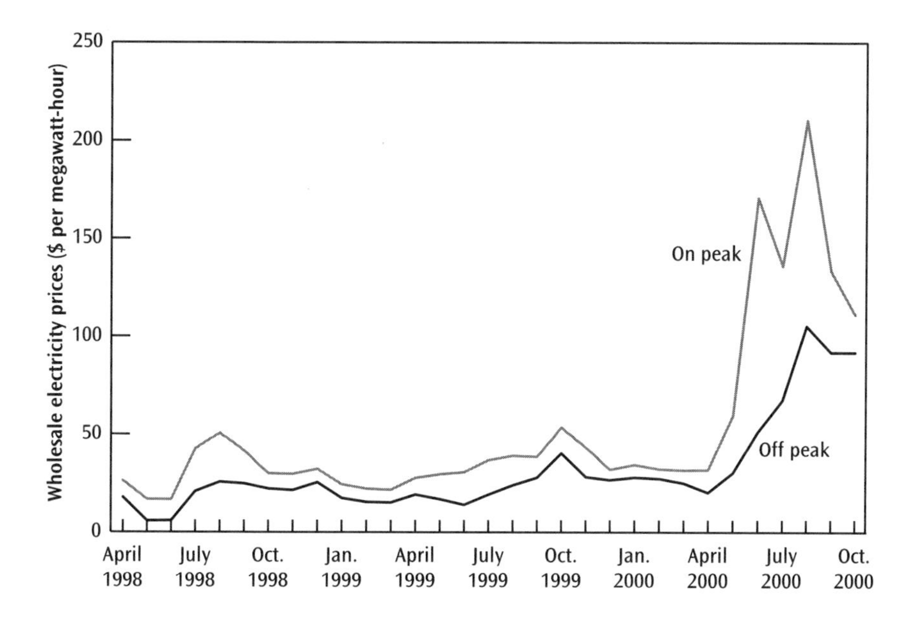

Figure 5-1

California Wholesale Electricity Prices, 1998–2000

Source: California ISO, "Event Log," http://www.caiso.com/docs/09003a6080/08/8a/09003a6080088aa7.pdf (accessed May 2001).

or generators (through government-ordered rebates) wanted to pay. As the utilities teetered closer to bankruptcy, power producers became less willing to sell them electricity. Prices in the winter of 2001 continued at levels about 10 times the 1999 average, despite winter being a typically off-peak season for electricity demand in California.

Alerts became much more frequent, with increasing power interruptions and some rolling blackouts in northern California. Citing a Federal Energy Regulatory Commission (FERC) order forcing it to "implement a $150 breakpoint" that, apparently, was below the price generators were willing to offer power, the California Power Exchange (PX) suspended operation on January 31, 2001. Within two weeks, a federal district judge forced generators to continue to sell power to the utilities, despite the utilities' failure to maintain an "approved credit rating." The court cited "emergency dispatch" provisions in the ISO's rules and the severe harm if generators refused to provide power in California

State and FERC Responses

As the financial status of the distribution utilities worsened, the state government reluctantly began to act, mostly by expanding its direct role in the industry. The legislature authorized the state to purchase power, finance plant construction and retrofitting, fund conservation programs, and sell power directly to retail customers. The governor further proposed that the state would apply cost-plus regulation to power supplied by plants the utilities kept. Partly to relieve utility debt, California Governor Gray Davis commenced negotiating deals to purchase the transmission lines owned by the utilities. To alleviate some of the pressure on the market created by caps on emissions of nitrogen oxides (see Chapter 15), California regulators lifted limits, allowing generators to emit nitrogen oxides at a price of $7.50 a pound, about five times historical levels but about one-fifth the price to

Figure 5-2

Staged Emergencies in California, April 1998 through October 2000

Source: California ISO, "Event Log," http://www.caiso.com/docs/09003a6080/08/8a/09003a6080088aa7.pdf (accessed May 2001).

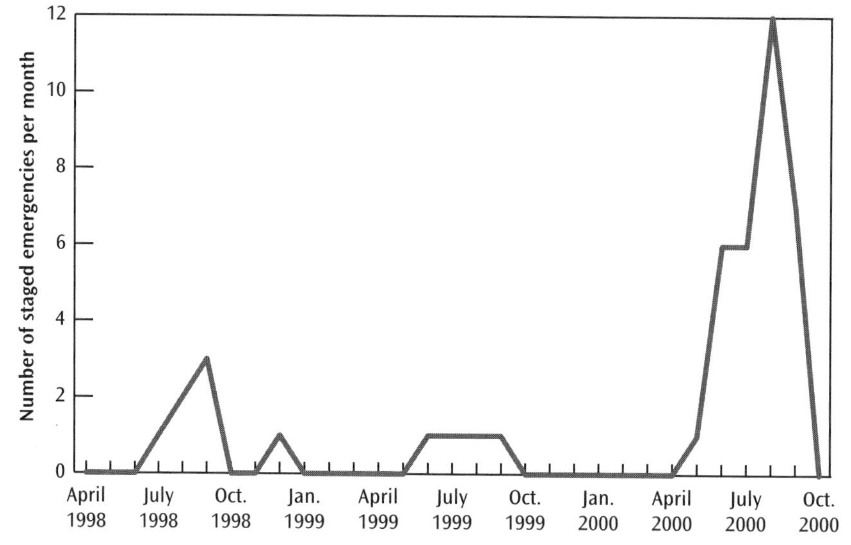

which they had risen during the crisis. The California Public Utility Commission decided that to cover the cost of electricity, it would bite the bullet and raise power rates—particularly to industrial and commercial customers—by 40%.

California and the federal government, primarily FERC, continued to dispute the degree to which generators had illegally violated federal statutes requiring the wholesale electricity prices be "just and reasonable." In March 2001, FERC reviewed wholesale power prices for excesses due to market power and, on the basis of a proxy market price of $273 per megawatt-hour, ordered $69 million in refunds for overcharges that occurred in January 2001 during emergency conditions. The California ISO disagreed, and it provided studies that estimated an overcharge during that month of $1.364 billion, almost 20 times FERC's estimate. The difference was based in part on the ISO's view that prices were too high all the time, in contrast to FERC's view at the time that prices could be too high only when capacity was pressed. For PG&E, these responses may have been too little, too late; it filed for Chapter 11 bankruptcy protection on April 6, 2001.

Subsequently, in April 2001, FERC issued an order requiring power producers to bid in available power to the ISO, and capped wholesale electricity prices in California during periods when unused reserve capacity for electricity generation fell below 7% of the power produced. Generators could get waivers to sell power at prices above the cap with appropriate justification. About two months later, FERC expanded this policy to set caps on wholesale power at all times, and throughout the Western United States rather than just in California. During times when reserves were low, the cap would be based on the cost of the most expensive gas-fired generator; at other times, the cap would be 85% of the price set during the most recent period of low reserves. The price caps are scheduled to remain until September 2002.

While FERC was taking a stronger role in the market, it continued to be embroiled in the dispute regarding the size of refunds due California. Under the auspices of FERC's chief administrative law judge, settlement talks between California and the generators took place in June and July 2001. The talks proved unsuccessful. California took the position that it was owed $8.9 billion, whereas the leading generation company proposed in the aggregate about $700 million. The chief judge noted in his July 12, 2001, settlement report that "while there are vast sums due for overcharges, there are even larger amounts owed to energy sellers by the CAISO [California Independent System Operator], the investor-owned utilities, and the state of California."

As the summer of 2001 began, reliability experts forecast that California might suffer as much as 260 hours of rolling blackouts. Those dire predictions were not borne out, as California got through the summer without a power outage, with August prices falling to $53 per megawatt-hour, about a third to a quarter of the preceding year's levels. Just as the blackouts themselves were likely the result of numerous factors (described below), the ability of the state to avoid serious problems this summer was also attributed to a variety of causes. Some of the credit goes to an unusually effective conservation campaign, reducing electricity consumption in California by more than 12% in comparison with the previous summer, according to the Natural Resources Defense Council. However, another major contributor was a summer that was milder than expected. Nevertheless, the state continued to walk away from its deregulation efforts. On September 20, 2001, the California Public Utility Commission voted to end giving electricity consumers the ability to choose their electricity suppliers.

Potential Culprits

The severity of the power crisis has inspired an ample supply of potential explanations. Were it possible to convert explanations into electricity, California's dilemma would disappear. We identify below a list of 10 potential culprits (not necessarily in order of significance):

- *Supply-and-demand imbalances:* The capacity to produce and deliver electricity to users in California failed to keep up with growth in demand.
- *Higher fuel costs:* Prices for power rose because the fuels used to produce it, particularly natural gas, became more expensive.
- *Supply-reducing regulations:* Rules restricting the construction of generation and emissions of particular pollutants reduced electricity supplies and raised generation costs.
- *Wholesale price regulation*: Limiting or threatening to cap the prices a generator could charge for power sold to the distribution utilities could have discouraged supply.
- *Retail price controls*: Holding down retail prices kept demand high during peak periods and caused distribution utilities to lose money when they had to purchase wholesale power at high prices.
- *Inframarginal rent transfer*: Moving from regulation to competition implied that when prices began to rise to cover the cost of the "marginal" firm, all of the suppliers were able to charge that high price, redistributing wealth from consumers to producers.
- *Absence of real-time metering*: The general absence of devices to measure power use at any given time prohibited setting power prices high during peak-use

periods, removing an important incentive to conserve power and reschedule use for times when electricity was more plentiful.

- *Bad luck–lack of long-term contracts*: Rules requiring distribution utilities to buy power from the Power Exchange kept them from insuring against high wholesale prices via long-term contracts.
- *Auction design*: Instituting an auction in which suppliers could get the highest accepted bid price could have created incentives to "game" the system.
- *Market power*: Through collective action or unilateral conduct, generators may have charged prices substantially above the competitive level.

In assessing these and any other explanations for California's woes, remember that restructuring in California worked well for more than two years and has worked reasonably well in other parts of the country and throughout the world (see Chapter 4). Any explanation that insists that opening electricity markets to competition necessarily leads to a California-like crisis is intrinsically dubious. Opening electricity markets may be a necessary condition for some—not all—of these potential explanations, but it is not sufficient.

Supply and Demand

The first three items on the list (supply-and-demand imbalances, higher fuel costs, and supply-reducing regulations) affected the overall supply-and-demand balance in California. They have little if anything to do with restructuring per se; their effects would have been felt even if the electricity markets in California were still regulated. Other items on the list, of course, could and likely did aggravate the effects from these supply-and-demand factors.

Consumption Overtaking Capacity

At the head of the list of causes for the crisis is that California hit a very hard wall when steadily growing demand, fueled by increased population and a strong economy, strained production capacity to the limit. During the 1990s, demand for electricity in California grew by more than 11% while capacity fell slightly. Population and economic growth in other Western states reduced electricity supplies that California might otherwise have imported. Exhausting the capacity to produce electricity would have led to higher prices, rolling blackouts, or perhaps both—even had California not adopted restructuring.

The supply crunch was exacerbated by a lack of rainfall in the Western United States, reducing hydroelectric power supplies in the summer of 2000 by more than a quarter below production in 1999. Along with generation constraints, limits on transmission have hampered the delivery of electricity into and within California to areas most in need. The combination of net capacity reductions and demand growth with the other attributes of electricity—very inelastic demand (abetted by rules and impediments to flexible retail pricing) and an inability to store electricity for use in times of shortage—is a sure recipe for significant increases in wholesale electricity prices and blackouts. Other factors described below may have made a bad situation worse, but had capacity remained plentiful relative to demand, electricity policy would have remained an obscure pursuit.

Higher Fuel Costs

The price for power rose in part because the fuels used to produce it, particularly natural gas, became far more expensive. According to FERC, natural gas prices nearly tripled in the Western United States during 2000. Among the portfolio of generation technologies used to supply electricity in California, natural gas–fired generators are the ones called upon to meet peak demands. As supplies get tight relative to demand, the generators called upon to meet power needs are less and less efficient. Consequently, when the price of natural gas rose, the cost of meeting peak demand increased, raising the price required to attract enough electricity to meet demand.

Supply-Reducing Regulations

An additional factor possibly adding to a supply-and-demand imbalance was environmental regulation. One possibility was that regulations limiting emissions of particular pollutants (see Chapter 15) reduced electricity supplies and raised generation costs. Many commentators on the California energy situation attribute the increase in electricity prices to the increase in the price of permits to emit nitrogen oxides. This explanation confuses effect with cause. The increase in prices of the permits is a reflection of, not a contributor to, the increase in electricity prices. As electricity became more expensive, the demand for permits rose, allowing those who owned permits to earn scarcity rents. If the permit program mattered, it did so as a quantity constraint, not as an input price.

A longer-run environmental constraint on supply in California could be so-called NIMBY ("not in my backyard") attitudes that impede power plant siting, approval, and construction (see Chapter 17). The California Energy Commission, the state agency with the authority to approve the construction of energy-related facilities, cites among the "critical issues" affecting siting "availability of emissions offsets, ... local agency and public opposition, [and] land use constraints." Some environmental advocates deny that environmental regulations have had much of an effect on power plant construction.

The important point is not to attribute blame. Environmental rules can and inherently should raise the prices of commodities such as electricity when pollution is a by-product. If not, customers will ignore environmental costs and think of power as too cheap, and thus use too much of it. Whether the price increase attributable to environmental regulations in California is above or below the value of the environmental benefits generated by those regulations is for the political institutions in that state to decide.

Regulation and Redistribution

The next 2 items on the top 10 list, wholesale price regulation (actual or threatened) and retail price controls, concern the ways in which price regulation may have caused or amplified the California crisis.

Wholesale Price Regulation

Calls to cap wholesale power prices have been made since the crisis began and are becoming more prominent in academic and political arenas. Beginning in December 2000, and continuing until June 2001 as was noted above, FERC issued a series of increasingly stringent controls. They may alleviate a politically undesirable redistribution of wealth from consumers to generators. However, limiting or threatening to cap prices could have discouraged supply. Even though these caps were "soft" (able to be waived) and at least five times the usual price of power in California before the crisis, they could and perhaps did impede supplies. The unwillingness of generators to abide by caps imposed in December 2000 led in part to the closing of the California Power Exchange in late January 2001.

Supply reductions with a wholesale cap ought not be surprising. Peak electricity can be an expensive commodity to supply. Plants that meet peak needs must cover their capital costs in much less time than that available for a plant operating continuously. If generators do not expect to recover these costs, they will not enter the market. However, if the wholesale power market is not competitive, then a price cap could increase supply (Chapter 9).

Retail Price Controls and Paying the Bill

If hitting the capacity wall was the primary cause of the crisis, holding down retail prices made matters worse. Low retail rates would keep demand high and discourage conservation that might have eased the stress on the power system. Utilities lost billions of dollars when they had to purchase wholesale power at prices five or more times the retail level to meet their legal obligations to serve the public. The potential for bankruptcy called into question their ability to pay, leading to a vicious circle in which wholesalers would raise prices to cover the risk of nonpayment.

The disastrous nature of keeping retail rates low when wholesale prices skyrocketed is obvious, but perhaps only in hindsight. Optimistic expectations when restructuring was enacted, supported by the first two years of the California electricity experience, might have encouraged regulators to think that the retail controls were not going to be binding. Such controls may have been viewed as protection against market power in retailing, as long as the incumbent distribution utilities retained a near monopoly. The retail price may have been held down as part of a political bargain, to redistribute some of the expected gains from restructuring back to consumers in the form of lower power prices. Now, the political problem is to distribute the wholesale bill among distribution company stockholders (bankruptcy), electricity customers (rate increases), California taxpayers (state-funded bailouts), and the generation companies (FERC-ordered refunds, court-mandated obligations to serve, and fractional debt repayment).

Redistribution from Consumers to Generators

One virtue of opening previously regulated markets to competition is that it makes prices, hence costs, visible. Peak costs were not visible because under regulation peak costs would be averaged in with the lower costs in off-peak periods. Markets operate differently. As prices rise to cover peak load generation, everyone in the market, not only the marginal firm, gets to charge the high price. This outcome is efficient in that higher prices send the right message about the value of conservation. The cost of drawing electricity from a low-cost baseload plant equals the cost of replacing that electricity from a high-cost peak-load plant.

But the transfer is likely to be politically upsetting, thus placing it on the list of potential contributors to the crisis. The initial symptom of the California crisis was not blackouts or bankruptcy but the political turmoil associated with higher retail rates in San Diego in the summer of 2000, during the three-month window in which its retail rates were not regulated. The long-term solution is for more generators to come online. Off-peak power prices would fall, perhaps down to operating costs, as peak-load sales produced only enough revenue to cover capital costs. (Resort hotels that charge cheap off-season rates and cover their expenses with high tourist-season prices illustrate what could happen.) The political will to wait for entry may be hard to come by, as new generation construction is not a quick process.

Market Design and Mechanics

Absence of Real-Time Metering

Because electricity cannot be stored, the cost of producing it and the value of conserving it are highly sensitive to the time when it is used. Electricity may cost 10 times as much to produce on a hot summer afternoon as it costs later that same evening. But standard meters, which tell only how much electricity one uses during a particular month, do not allow prices to be charged according to the time the power was used. As long as consumers pay the same price for electricity regardless of when they use it, they will not reap these incentives from cutting back power use or, to put it perhaps negatively, pay the full price for power when they use it. Consequently, they will use too much power and have too little incentive to conserve.

Real-time meters would allow such charges, encouraging consumers to conserve and reschedule use for off-peak times. Suppose turning up a thermostat a few degrees would reduce the use of a 5-kilowatt air conditioner by two hours a day during peak-demand afternoons. A consumer who typically pays $300 per megawatt-hour (MWh) during those hours would reduce his or her electric bill by about $90 during that month but would save only $9 at an off-peak (or average) price of $30 per MWh. A necessary step to getting users to see prices that reflect costs in real-time prices would be the widespread use of real-time meters.

If retail prices are not otherwise held constant by regulation, additional real-time metering would reduce demand for electricity during peak periods. But the desirability of real-time metering need not imply policies to increase their use (e.g., through subsidies or regulatory order). If a utility is selling power at $30 per MWh while paying $300 per MWh for it, one might think that it already has a strong incentive to promote real-time pricing, including paying its customers to install real-time meters.

From an economic perspective, the key question is whether all customers benefit when any given customer installs a real-time meter. The strongest argument for doing so is when the alternative to installing real-time meters is not merely financial—the distribution utility loses money—but involves actual blackouts. If A's purchase of a real-time meter reduces the chance that B will be blacked out, then B benefits as well. Policies to subsidize real-time metering can stand in the place of payments B would be willing to make to A to get A to install the meter and make blackouts less likely. A second justification is that because blackouts cover entire areas, an individual consumer cannot easily buy his or her way out of blackouts by

Spot Markets and Contract Sales

A fundamental distinction in how electricity, and virtually any other commodity, may be purchased is between spot sales and contracts. *Spot* sales refer to purchases made "on the spot" (i.e., right at the moment). One pays the going price at the time. Electricity purchased under *contract* is typically bought in advance of the time of delivery.

Electricity is not the only commodity traded over both contracts and spot sales. *Futures markets* offer businesses, investors, and speculators the opportunity to set prices in advance for a wide range of agricultural products, minerals, and financial assets. One can think of hiring labor in spot and contract terms, where a spot purchase would be akin to hiring a temporary employee for a day at the going rate, whereas a permanent employee on salary resembles a contract.

Contracts provide two advantages over spot sales. First, they may reduce the costs of arranging for sales, because both buyers and sellers have time before delivery to find the best price and negotiate the most favorable terms and conditions. Second, as we note, they provide insurance, in that buyers and sellers know what price they will get in advance, and thus need not worry that prices may unexpectedly rise (the buyers' concern) or fall (the sellers' concern). The disadvantages of contracting are the flip side of the advantages—buyers and sellers may lack the flexibility to adapt to changes in market conditions. They may also come out on the losing end, for example, if an electricity marketer contracts in advance to buy energy at one price, but when the time comes to sell the energy, the price has fallen.

In practice, the distinction between contract and spot sales is one of degree as much as one of kind. Buyers and sellers of electricity will typically employ a mix of spot sales and contracts in their transactions. In addition, contracts vary in a number of important dimensions. They can vary in the time between they are signed and the exchange takes place, and the conditions for which a buyer or seller can "get out" of the contract (e.g., by a seller refusing to deliver the product and giving the buyer its money back). Rather than requiring purchase, contracts also may give parties the *option* to buy (a *call option*) or sell (a *put option*) at a given price within a specified length of time. Ultimately, *vertical integration,* that is, having the buyer and supplier (e.g., generator and transmission company) in the same company, can be thought of as a surrogate for a long-term contract, in which the terms are implemented by putting "everything under the same roof."

agreeing to install a real-time meter and pay a high peak price in exchange for a steady flow of power.

The same arguments may apply to substitutes for real-time pricing, that is, programs to reduce electricity use during peak periods, such as interruptible service contracts or private demand-side management programs. In addition, we ought not ignore the possibility that the main problem in California was not merely that peak retail prices were too low but that the overall average price was set too low by regulation. If so, prices should be increased, whether or not real-time meters are installed.

Bad Luck: Lack of Long-Term Contracts

An idiosyncratic aspect of the California experience was that the distribution utilities were discouraged from hedging against high wholesale price increases through long-term contracts with generators. If the dramatic increase in wholesale prices in the summer of 2000 had not been expected with a very high probability (as the previous two years of low prices would suggest), the utilities might have been able to obtain favorable long-term supply contracts at fairly low prices. Had they done so, they and the State of California would not be suffering the financial strains associated with bankruptcies and bailouts.

Yet it is important not to let hindsight exaggerate the potential benefits of these contracts going forward. Long-term contracting is essentially purchasing insurance from generators against high prices in the future. (For more on the features of spot and contract markets, see the box.) Looking at the absence of such contracts in retrospect is not very different from observing that someone whose house just burned down should have bought fire insurance.

The salient question would be whether the absence of long-term contracts hurts electricity production. Long-term contracts, like other forms of insurance, are financial, not "real" in the sense of producing more electricity as such—fire insurance does not construct houses. In addition, like other forms of insurance, long-term contracts can lead to *moral hazard*—that is, being less careful after reducing exposure to risk. In electricity, long-term contracting could encourage greater consumption at a lower contract price and discourage conservation.

Hindsight might also exaggerate expected price savings. The price of the long-term contract, like an insurance premium, will reflect expected costs down the road. Price increases in the summer of 2000 and beyond may have been so surprising that long-term contracts might have resulted in huge savings for the distribution utilities (and perhaps substantial losses for the generation companies who would have had to honor them). But going forward, one would not expect the price of a long-term contract to differ greatly from the price that generators expect selling short term in the future.

A final potential problem with long-term contracting arises when the contracts are between distribution utilities and generation companies. A leading concern in electricity policy has been to promote independence between the regulated monopoly "wire" sectors of the industry, and the competitive generation (see Chapters 6 and 7). As long as the distribution utilities are also the retail marketers of electricity, as they largely have been in California, long-term contracts could reestablish links that restructuring policy has been striving to break.

Nevertheless, the inability to spread risk efficiently between generators and distributors could discourage entry by buyers or sellers. If fire insurance were impossible, people would likely buy fewer houses. Wholesale price volatility in California has undoubtedly increased the importance of efficient risk sharing. Moreover, because financial insolvency has real effects in a market where balance sheets constrain liquidity, the inability to insure against high wholesale prices via lock-in contracts has had real costs.

Auction Design

The California Power Exchange ran an auction in which, for every hour of the day, each generator could specify up to 16 prices, along with the amounts of electricity it would sell at those prices. For example, a generator could offer 200 megawatts (MW) of power at $20 per MWh and an additional 100 MW at $40 per MWh. The PX would find the price at which the amount generators offered equaled the quantity demanded. Each generator would then get that market-clearing price for all the supplies that were taken in the auction.

At first glance, the design seems reasonable. Because each generator gets the market-clearing price, they would appear to have little reason to act strategically. They would have no incentive to put in bids above their costs of electing to provide service at the particular hour. The supply curve produced by combining the bids would be akin to a supply curve in a competitive market.

The rule that all bidders get the market-clearing price seems to mimic a textbook competitive market but differs from real-world markets in an important respect. This practice may change the incentives of the sellers. In the California electricity auctions, the implication is that generators may have an incentive to bid in a little bit of power at a very high price. If a generator's bid for that power is not taken, it does not lose much—only the profits from the small amount of sales. If the bid is taken, however, the generator could reap a windfall, in that it receives that high price on all of its output, not just the amount bid in at the high price.

A helpful analogy is a car dealership where several cars of the same model are on the lot. The dealer considers putting a very high sticker price on one of the cars, on the chance that someone might show up so desperate for a car that he or she would be willing to pay the high price. This tactic seems unprofitable. The car will likely remain unsold, and the only benefit depends on the slim chance that a desperate customer unwilling to shop around would appear to buy that one car. But suppose that customers may be desperate, and if the car were to sell at this high price, the dealership would be able to go back to customers who had bought cars there previously and make them pay the high price as well. The car with the high sticker price could go unsold, but the expected benefits from setting the high price are now considerably greater.

The California electricity market may be like the second scenario, with desperate, inelastic demanders and the ability to charge all buyers—not only the desperate marginal buyer—the high price. The generators then could have an incentive to bid in their last megawatt-hour at a very high price. The only cost is the forgone profits from selling that last megawatt-hour, whereas the expected gain is that all the power would be sold at that high price if it ends up being taken.

This possibility is hardly assured. Each member of a set of bidders may prefer to bid in all of its supply at cost and let someone else take on the burden of setting the high bid—so no one does it. In addition, bidding rules would have to require that generators offer a minimum quantity at any price, but a quantity small enough so the cost of it going unsold is reasonably small.

Some have suggested that the state could reduce electricity expenses by dropping the rule that all bidders get the market-clearing price. However, this would be unlikely to produce benefits and likely to make matters worse. First, generators would have an incentive to raise their bids above their costs, because what they make would then depend on what they bid. This, in and of itself, would tend to reduce the amount of electricity that generators are willing to offer at any given price, thus leading to reduced supplies and higher prices for the "last" or marginal megawatt-hour of energy necessary to meet demand. In turn, generators would make offers on the basis of this higher expected price for the marginal megawatt-hour, countering the rationale for paying generators what they bid rather than the market price.

Long-term contracting would reduce the incentive to place high bids by taking some output off the market. Forcing suppliers to offer all power at the same price, or at least increase the minimum quantity that a generator would have to offer at any price at which it bids, would increase the cost of not being able to sell that power if that bid were not accepted. One might also ask whether the reach of grid operators should be extended into the business of organizing electricity markets, rather than letting buyers and sellers contract as they choose (see Chapters 10 and 11).

Market Power—Last but Perhaps Not Least

Perhaps the most controversial claim regarding the California situation is that it is the result of the last item on the list, the exercise of market power by generators that supply power to California. To differentiate this issue from those associated with exploiting auction design, by *market power* we mean the withholding of output to raise prices. In this situation, this would refer to intentional outages or supply reductions by generators that supply power into California.

Such concerns predated the summer 2000 crisis. In March 2000, the Market Surveillance Committee of the California ISO found that the state's electricity markets had not been "workably competitive" during the summers of 1998 and 1999, when prices were estimated to have exceeded competitive levels by less than 20%, rather than the fivefold increases in price observed after June 2000. Ironically, if generators exercised market power before June 2000, something else was responsible for causing prices to increase by factors of five or more above prior prevailing levels. That does not mean that market power played no role in wholesale electricity price increases from June 2000 onward. But it does call into question any view that deregulation of power markets ipso facto leads to the exercise of market power.

Because market power is likely to be a generic issue facing electricity policymakers across the country, we save much of our analysis of the issue for Chapter 9. In California specifically, features of its markets and policies could have made any exercise of power more attractive:

- Keeping retail electricity prices regulated insulated consumers from higher prices, essentially fixing demand and increasing the profits from raising prices.
- More widespread use of real-time meters would also make demand more sensitive to price, forcing a generator to reduce output by a greater amount to make a higher price stick.
- The general lack of long-term contracts in buying electricity from generators meant that if they cut back output to raise price, they would get to charge the high price for all their sales, not just those not already fixed by prior agreement.
- California's auction design, allowing multiple combinations of prices and bids, may have given generators an incentive to offer a small amount of electricity at very high prices, with the possibility that if that electricity were needed, the generator would get that high price on all of its sales.
- Separating markets for power from markets for ancillary services, though necessary to keep the electricity system operating smoothly (Chapters 2 and 10), could reduce the ability of firms to compete in both markets at once.

Whether this potential for the exercise of market power in California was realized remains a subject for dispute. Some claims, and at least one case, have been filed suggesting that generation companies colluded to raise electricity prices in California. Such accusations remain ultimately a matter of specific evidence. In principle, fixing prices of electricity in California may have been relatively easy in comparison with other commodities, because it is basically an identical good sold by all in a single market. But the sheer number of competitors in the California wholesale market would seem to make collusion difficult. Moreover, because cartels are illegal, they have a strong interest in avoiding publicity and scrutiny, and presumably would not want to restrict supply to the point of creating blackouts, putting major utilities into bankruptcy, and draining the state's treasury.

A more likely possibility is that generators unilaterally had the ability and incentive to withhold output and raise prices. But each seller may have found it unilaterally worthwhile to reduce output, especially when retail regulation and the absence of real-time metering would have made demand largely insensitive to price. Whether outages were designed to raise prices remains under regulatory scrutiny. As we discuss in Chapter 9, some studies find that electricity prices were above the average variable costs of generating power, but these findings need to be interpreted with care. Peak-period prices would normally cover not only variable costs but capital costs as well. Moreover, the prices charged may have been inflated to compensate for the possibility that bankrupt utilities might not have been able to pay their bills.

If unilateral market power is a problem, in California or elsewhere, the antitrust laws are not an effective remedy. If a firm legally acquired its market position, as seems the case with the generators selling in California, they can charge under the law whatever they want. If an electricity market is insufficiently competitive, and the high prices are not the fault of market conditions, regulatory distortions, or auction rules, then one of two difficult remedies is necessary. The first would be an order to force generation companies to divest some of their plants to deconcentrate the market. Such a divestiture would be unprecedented in an industry as nominally unconcentrated as electricity generation. Moreover, new legislation might be necessary to give FERC or whichever entity that would issue such an order the authority to do so. If that strategy fails (e.g., because reasonable scale economies in generation set limits on how much divesting the industry can take), the alternative would be to regulate wholesale electricity prices. Whether such regulation is temporary or comes to be a permanent fixture of restructured electricity markets remains to be seen.

PART II

Current Policy Issues

CHAPTER

6

Competition in Energy, Regulation of Wires

For many industries where regulation has been replaced with competition, such as trucking or banking, much if not most of the industry has been largely freed from continued regulatory oversight. But in some sectors—telecommunications, for example—the process of deregulation has been only partial, with continued regulation of some segments. If the industry could just be totally deregulated, the policy task would be much simpler.

When deregulation is partial, it becomes more complicated. Policymakers need to establish the boundary between the regulated and deregulated sectors of the industry, how to manage the relations between the sectors, and how best to continue regulation where necessary. This chapter includes a look at why electricity is one industry that in part—that is, the wires that carry electricity from the power plant to the home, office, and factory—will remain regulated. Paradoxically, regulating less of the electricity industry could make regulation itself more complex. Technologies in the offing that will allow generation at the user's location could reduce the need to regulate the industry altogether.

Since the late 1970s, deregulation has been a dominant policy watchword. The past 20 years have seen the elimination of price and entry regulations in a number of industries, notably airlines, banking, and long-distance telephone service. One major lesson from this experience, particularly from telecommunications, is that the transition from having the government set prices in an industry to letting the market set them goes much more easily if it is complete rather than partial.

Unfortunately, not all segments of the electricity industry are equally conducive to competition. Where to draw the line between competition and regulation is a matter of judgment, which is informed very much by one's empirical and ideological beliefs regarding their relative merits. Skeptics regarding competition question the degree to which firms in an industry will compete with each other and the benefits such competition brings to consumers and the economy. Skeptics regarding regulation question the ability and inclination of the government to set prices to protect the public rather than special interests, and to preserve appropriate incentives to cut costs, improve quality, and stimulate innovation.

Nevertheless, a broad consensus has emerged that a line should be drawn between the potentially competitive segments of the electricity industry and those that should remain regulated for the foreseeable future. To oversimplify a bit, that consensus would retain regulation of the "wires" side of the business—local distribution and long-distance transmission—while extending competition to the "energy" side of the business—the generation and marketing of electricity. This consensus has been reflected in legislative and regulatory decisions implementing electricity competition in electricity markets in the many states that have taken that step. It also serves as the basis for the Federal Energy Regulatory Commission's (FERC) endorsement and, to some degree, requirement that control of transmission be functionally separated from ownership of generation through independent system operators (ISOs) and regional transmission organizations (RTOs).

Maintaining regulation and competition simultaneously in electricity (or any other industry) requires solving some difficult problems, including

- defining the boundary between the regulated and competitive sectors,
- redesigning regulation in moving from setting the price of electricity itself to setting the prices of just the regulated components,
- managing dealings and corporate organizations that cross the boundary between the regulated and competitive sectors, and
- deciding how the regulated or competitive sectors should be responsible for maintaining the economic and technical reliability of the overall industry.

Solving these problems first requires an understanding of why there is likely to be a regulated sector in electricity for some time to come. This first entails clarifying the general principles that determine when regulation is warranted. Applying these principles to the electricity industry explains why its wire portions—transmission and distribution—are likely to remain regulated while generation and marketing become competitive.

Why We Regulate Prices

The fundamental organizing principle in the U.S. economy is a reliance on market competition to set prices. Ideally, firms, in their quest for profits, seek to provide consumers with the things they want to buy and to produce them at the lowest possible cost. As part of this process, firms will want to leapfrog each other in coming up with innovative goods and services and in devising the technologies to make and deliver them. Competition, however, leads firms to undercut each other on prices, until consumers need pay only what it takes to cover the costs of producing what they buy. The result is an economy in which people generally can choose whatever they want among an ever-expanding set of options, and they have to pay only what it costs to make it.

This textbook picture need not match the real world, for a variety of reasons. In some markets, such as insurance or lending, buyers and sellers may not be able to share pertinent information about costs and benefits. In extreme cases, this *asymmetric information* problem can cause a market to disappear—for example, when people who know they tend to be healthy opt out of buying insurance, leaving only those so prone to be ill that the cost of insurance is prohibitive. Another problem, more pertinent to electricity policy, is that one person's choices may harm a third party; my use of electricity, for example, may increase air pollution over a broad region, essentially imposing an extra cost on society that is not reflected in the market price of electricity (see Chapter 15).

The most important aspect of a market that pertains to price regulation is whether there is sufficient competition to drive prices down to a level reasonably close to cost. Without such competition, a firm may feel it can raise prices without the fear that customers will flock to other sellers that are either in the business already or would enter the market in response to the profit opportunity created by these high market prices. The extreme case where this is a concern is if there is only one seller in a market—a *monopoly*—that faces no potential competition from the outside (see box).

Economists refer to the ability to raise price, insulated from competition, as *market power* or *monopoly power*. The mere existence of detectable market power, however, usually does not warrant having the government regulate the price. Just as markets diverge from ideal outcomes, so do regulators. They may not be able to get the information necessary to determine a reasonable price, especially if the product's costs or characteristics are changing rapidly. Moreover, politics being what it is, they may tend to "regulate" more in the interest of firms with market power, which have a direct stake in the outcome, and with less regard for consumers, who might not have the wherewithal to organize effectively.

Looking at where and when we regulate and where and when we do not, suggests three rough criteria for when we choose regulation over markets: if the seller is a "natural" monopoly with no significant competitors, the service is one that most

What Is a Monopoly?

A monopoly is a market with only one seller. By being the sole supplier of an item and facing no competition, a monopolist can select the price that is most profitable for itself. For many people, the harm of monopoly is primarily the transfer of wealth to a monopolist from the buyers who have to pay the monopolist's prices. Economists emphasize the inefficiency that follows when a monopolist raises its price, because to do so it generally must reduce overall output in its market below what would be sold at lower, competitive prices.

Monopolies can be public when the government is the sole supplier; the U.S. Postal Service and municipal transit, water, and electricity distribution companies are examples. In the United States, though, most monopolies are privately owned. Whether private or public, monopolies are sometimes created by the government through laws and rules that limit entry and competition. Firms may attempt to create monopolies through collusion, but antitrust laws exist to prevent such conduct. When technology and capital costs imply that one firm can provide a good or service at less expense than two or more firms—that is, when a natural monopoly exists—those markets may come to be monopolized as well. In all of these cases, the government may act, sometimes for good reason, to regulate the prices these monopolies can charge.

The presence of natural-monopoly conditions, however, does not necessarily imply the absence of competition. If the cost advantages for a single firm are not great or if competition is not especially intense, multiple competitors may be able to survive in a market. Alternatively, a firm with a natural monopoly may feel pressure from potential entrants that want to replace it as the monopolist in that market. Finally, the monopolist may face competition from suppliers of other commodities that can be substituted for it. For example, an electric utility may face competition from other fuels (e.g., oil or natural gas) for some uses (e.g., home heating).

Source: Timothy J. Brennan and others, *A Shock to the System: Restructuring America's Electricity Industry* (Washington, DC: Resources for the Future, 1996), p. 65.

consumers value greatly, and the technology for providing the service is relatively stable.

The Seller Is a Natural Monopoly with No Significant Competitors. A *natural monopoly* refers to a setting in which one provider can offer a product or service to all comers at a lower cost than if the market were divided among two or more sellers. Where this holds, the average cost of production falls, the bigger is the size of the operation—exhibiting what are called "economies of scale" (see the box titled "Economies of Scope and Scale" in Chapter 7). The classic example would be an industry for which a factory is expensive to build, but once built can supply enough to meet demand with relatively low actual production costs.

A natural monopoly in theory is generally necessary to produce an actual monopoly in practice. Without a natural monopoly, a market can support at least two or more competitors. In virtually all markets where we regulate to control market power, there is only one significant provider of the service, indicating a view that even imperfect competition is preferable to regulation for getting cost-based prices and encouraging innovation. But even if the technical conditions for a natural monopoly are met, multiple suppliers may compete with each other (e.g., by offering distinctive but substitutable products). Therefore, we need to go beyond those technical conditions and examine actual competitive conditions before concluding that competition in a market cannot work.

The Service Is Valued Greatly. One can imagine lots of monopolies in practice—for example, the only bowling alley in a small town. Perhaps because the bowling alley lacks any direct competition, it can charge prices significantly above its costs. But, judging from experience, we would not bother to regulate the bowling alley. The costs of regulating—setting up an agency, gathering cost information, and enforcing price controls—exceed the benefits of reducing the price of a service that most people do not regard as crucial. Only services such as electricity, telephones, water, and natural gas delivery merit bringing in the public sector to prescribe prices, or in some cases to provide the service itself (e.g., municipally owned power companies; see Chapter 13).

The Technology Is Stable. Regulation is inherently slow. It takes time to propose pricing rules, get public comment, make decisions, have them reviewed by the courts, implement them, and enforce compliance. For the task to achieve some semblance of accuracy, the nature of the regulated product and the cost of producing it must be relatively stable. When technology is rapidly changing, neither condition may hold.

In such cases, even if we have natural monopolies that provide important products, we may leave pricing and product decisions to the market, such as it is, rather than the public sector. Computer software is an interesting example. It has the natural-monopoly characteristics of high fixed costs and low costs of serving additional users, along with the further effect that most people want to be on the "standard" system that almost everyone else is using. It is also a product crucial to the economy. But the constant stream of costly software investment and innovation should make even the least skeptical potential regulator think twice.

Deregulating Marketing and Generation

Electricity is undeniably crucial to the economy. The roughly 3% of the gross domestic product we spend on it—itself a huge number—understates the degree to which our households, offices, shops, and factories depend on it. Moreover, the technology for producing electricity is, relative to other industries, relatively stable (albeit to varying degrees).

The crucial factor in seeing where we might need to regulate electricity, and where we can open up to competition, is whether particular segments of the industry are natural monopolies unlikely to face competition. The marketing of electricity itself clearly is not a natural-monopoly service. Many firms could provide retailing and billing services, just as we see in many other markets. However, the "value added" in these marketing services themselves is probably not large enough to warrant opening these marketing services to competition, were nothing else in the sector to change.

Rather, the impetus to open markets in electricity comes largely from the potential for competition among generators. As was noted in Chapter 2, some of this potential for competition comes about because of technologies that allow for cost-effective generation at a smaller scale, making it feasible for more separate firms to operate profitably at competitive prices. An important contribution was the realization, following the experience with the independent power producers created by the Public Utility Regulatory Policies Act (PURPA), that the generation sector was not so tied to the wire sectors of the industry—local distribution and long-distance transmission. (Whether the generation industry will be adequately competitive, particularly during peak periods, is a subject broached by the alleged role of market power in the 2000–2001 California electricity crisis, discussed in Chapters 5 and 9).

That tie would not matter if the wire sides of the electricity industry were themselves subject to competition. But as was noted above and is discussed below, most observers regard distribution and transmission as natural monopolies. Until *distributed generation*—the ability of users to produce electricity on their premises—becomes more economical, distribution and transmission companies are likely to retain market power. As a consequence, the feasibility of competition among generators depends not only on the absence of scale economies that would otherwise have prohibited much entry into that sector. It also depends on the ability of generators to compete independently without all being part of the same company that owns the transmission grid and distribution system. Only with experience with mixing independent power producers with utility-owned transmission networks have we been able to introduce competition in generation after decades of regulation. Why transmission and distribution have economies of scale that preclude competition depends on important physical properties of electricity itself.

The Distribution Monopoly

The first reason local distribution is likely to remain a monopoly comes right out of economics textbooks. Local power distribution requires an extensive network of copper wires, poles and conduits, and transformers. Once a firm has built these facilities to serve a particular area, they can typically distribute all of the electricity that households and commercial enterprises in that area might want to use. For another firm to compete, it would have to make a huge, largely unrecoverable investment to duplicate that network, yet would offer no additional capacity to serve the market. That investment would be profitable only if competition between the two firms were so weak that the price of distribution would remain high enough to cover the costs of two redundant systems. (That price might even need to exceed the price an unfettered monopoly would charge.)

The bottom line is that no firm is likely to try to compete against an established distribution system. It would be unreasonable for the new firm to think that the price would remain high enough to recover its costs while competing effectively against the established firm. The lack of interest in building overlapping electricity grids during most of the past century confirms this proposition. In addition, the other wire-based networks offered by other companies—local telephone companies and cable television providers—are not likely to enter the power distribution business. Their systems are not designed to carry electrical energy at levels used by homes and businesses. The predictable result is that consumers will have only one distribution company from which to get their power.

This story is not unique to local electricity distribution. The natural-monopoly rationale for major physical delivery systems explains why most of us can choose from only one local telephone company, cable provider, water pipe system, and natural gas network. In some of these cases, competition may be becoming an alternative to traditional regulation. For example, wireless phones may compete with wire-based local phone systems, and telephone networks and cable systems may be able to compete in each other's markets for voice calling, television watching, and World Wide Web surfing. But because electricity is crucial, the technology for distribution is relatively stable, and no one is likely to enter or compete with established local networks, regulation is likely to remain the preferred means for ensuring that the prices generators and consumers pay for distribution remain reasonable.

A second rationale for continued monopoly regulation of local power distribution depends on a fundamental and important idiosyncratic property of electricity. Because electricity cannot be economically bottled up for future use to a significant degree, and because most uses of electricity are not deferrable, the amount people want to use at any given time has to equal the amount that generators pour into the system.

Some central authority needs to maintain this equality, known as *load balancing*, to keep the system operating in the face of variations in use and unexpected breakdowns. One way this was handled before electricity markets were opened to competition was to give the local distribution company authority over generation. Often, this authority was exercised through vertical integration, in which the distribution monopoly owned all the generators supplying it with power. In some cases, this authority was ceded to a regional power pool operator, which took responsibility for balancing generation and demand for a collection of geographically contiguous and interconnected distribution companies. This includes the ability to tell generators when they can and cannot come online—known as *dispatch* (see Chapter 2)—and to adjust the output of those generators that can most easily be changed while online—known, perhaps confusingly in this context, as *regulation*.

One of the major complexities in extending competition in electricity involves the degree to which load balancing will remain a function of the local distribution monopoly. Will the local distribution company (or transmission provider; see below) have to provide power "regulation"? Will it have to continue to provide centralized dispatch? Or can the responsibility to maintain an equal supply of and demand for power be put on the generators and customers themselves? We will return to these questions in Chapters 7 and 10.

The Transmission Monopoly

Local physical delivery systems often have natural-monopoly characteristics, but this is less true at long distances. The task of moving vast amounts of something from one place to another might be divisible among numerous providers, which could then compete with each other, eliminating the need for regulation. The most compelling example of this is in the telephone industry. Although the local telephone service monopoly is only beginning to face serious challenges, long-distance telephone service has seen significant competition for more than 20 years.

Applying this experience of the telecommunications industry to electricity would suggest that transmission could be subject to the same kind of competition we see among AT&T, MCI/Worldcom, Sprint, and other long-distance telephone companies. Unfortunately, however, electrons are not as cooperative as telephone calls and data traffic. Telephone calls and data can be routed through switches that specify which paths they can take through the overall communications network. (One could think of a telephone number as a set of switching instructions.) These messages can follow routes that have been specified in advance via contracts.

As electricity goes from a generator to a local distribution network or power user, however, it goes through all the wires that connect the former to the latter. With available technology, switching is prohibitively costly. Switching would not be necessary if there were only one line between any generator and a distribution network. However—mostly to promote system reliability and efficiency—the transmission grid is a complex, heavily interconnected set of high-voltage pathways between literally thousands of generators and destinations. This interconnection creates a wide variety of ways in which power can go from those who make it to those who use it at any given time.

These extensive interconnections, and the propensity of electricity to take all paths from origin to destination, creates what is called *parallel flow* or *loop flow*. Loop flow creates some thorny pricing problems, which we describe in other sections of the book. But the important idea here is that it tends to make an interconnected grid—often called an *intertie*—a single entity for economic purposes. Figure 6-1 illustrates this situation. If transmission utility 1 agrees to transmit power from generator 1 at point A to customer 1 at point B, some of that power will flow over transmission utility 2's interconnected lines. If transmission utility 2 increases the capacity of its lines, the benefits will go not just to it but to transmission utility 1 and any other transmission utility that owns a piece of the overall grid. It is as if a decision by MCI to make its network bigger would reduce AT&T's cost of carrying a phone call.

Loop flow effectively keeps interconnected transmission companies from each making their own pricing, marketing, and capacity decisions. Because each firm's decisions affect the capacity of others to transmit power, efficient operation of the

Figure 6-1

Simple Diagram of an Electricity Network

Source: Timothy J. Brennan and others, *A Shock to the System: Restructuring America's Electricity Industry* (Washington, DC: Resources for the Future, 1996), p. 77.

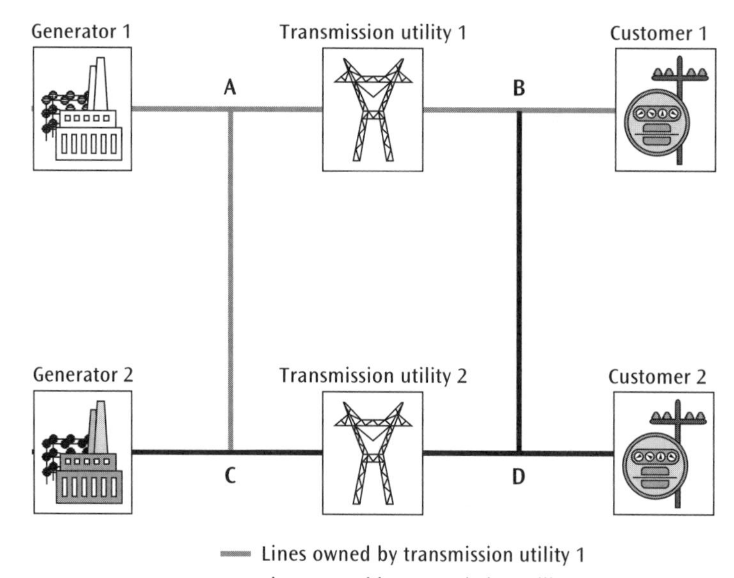

transmission grid requires cooperative management rather than competition among all its owners. In effect, it becomes a monopoly.

When electricity was provided largely by regulated franchised utilities, cooperative management was achieved through a variety of formal and informal power pools and reliability councils. Costs were covered in the regulated rates consumers paid for electricity. With the coming of open, competitive markets in generation, however, transmission becomes a separate function, requiring separate rules regarding access, ownership, and pricing. Because competition among independent transmission companies is not likely to evolve, transmission, like distribution, is likely to remain a separately regulated service for some time to come.

Does Less Regulation Mean More Regulating?

The natural expectation is that increasing competition in an industry decreases regulation. Although the scope of the sector that is regulated will fall, the effort that goes into regulation may rise. Before restructuring, regulators had a relatively simple and direct objective: Set the price of electricity at the level necessary to ensure that regulated industries cover their costs. Deregulation of power means competition can take its traditional role in setting prices and providing incentives for controlling costs and meeting consumer needs.

However, deregulation also means that a host of policies need to be put in place to ensure that remaining regulated transmission and distribution monopolies operate efficiently and do nothing to forestall competition among generators (Chapters 7 and 8). These include not only familiar policies to separate control of the wires from ownership of generation—such as creating ISOs or RTOs—but also the possible need to get the distribution companies out of the competitive business of retailing electricity directly to end users. In addition, regulators may need to take a more active role in ensuring that the power system remains reliable (Chapters 10 and 11). As was noted above and is discussed below, until on-site "distributed generation" becomes a widespread, economical, and reliable alternative to electricity delivered from distant large plants, we should not be surprised if restructuring increases the budgets, staffs, and dockets of state public utility commissions.

We might also ask whether much of the benefits of restructuring could be achieved by opening wholesale markets to competition, combined with better incentive-based regulation at the retail level (Chapter 8). Suppose utilities were regulated in such a way that they received a given price for electricity sales, where that price was not determined directly by the cost they incurred to produce power. They then would have an incentive to obtain power from the least-cost supplier, whether they generated it themselves or acquired it from an independent producer. We would still need to ensure that the owners of a transportation grid did not discriminate against unaffiliated producers in *wheeling* electricity to other utilities for resale to households, businesses, or industries. But the benefits from managing a smaller interface between regulated and competitive sectors could—we do not say "would"—exceed the costs from forgone opportunities from competition in electricity at the retail level.

Distributed Generation: Could Wires Lose Their Market Power?

Although competition among transmission networks is unlikely, technological change may break the transmission and distribution monopolies. This potential source of competition, as was noted above, goes by the name of *distributed generation*. Distributed generation refers to having power produced on one's premises—the home, store, office building, or factory—rather than produced off site and delivered via transmission and distribution lines. If distributed-generation technologies could produce power to meet demand at prices comparable to the cost of the generators and grids we use today, transmission and distribution network owners would face the likelihood that substantial numbers of customers would leave their systems if they tried to raise prices. Distributed generation could thus offer the competition necessary to limit the market power of these networks.

For a long time, the standard image of distributed generation was the rooftop solar collector that would convert sunlight to electricity. That technology, however, remains expensive, particularly when one factors in the cost of batteries necessary to store electricity for use at night. Some factories use the heat produced in their manufacturing processes to generate electricity on site. This *cogeneration* has been taking place for decades; in fact, cogenerators were among the first independent power producers to make electricity available for general use. However, the production patterns of cogeneration typically also do not match the use pattern over time—the "load profile"—of the power producer, implying a continuing need to remain connected to the distribution and transmission system.

A promising new technology for distributed generation, however, involves the use of fuel cells. At this writing, some firms have developed fuel cell generators that can provide power at prices comparable to the average cost paid for electricity in the United States today, about 7 cents per kilowatt-hour. This technology may not yet be suitable for most households. It may be some time before the technology can match the typical home or office load profile and before it is reliable enough for users to confidently cut off access to generators. However, regulators, policymakers, and certainly transmission, distribution, and power providers all will be keeping an eye on these developments.

CHAPTER 7

Vertical Restructuring

Much of the effort to eliminate price regulation during the past several decades in the United States has involved more or less complete deregulation throughout the entire sector. Electricity is an exception. Policymakers are opening power markets to competition, but local distribution and long-distance transmission are unlikely to be deregulated anytime soon.

Electricity is not the first deregulated industry to be split into regulated and competitive sectors. The telecommunications industry has seen a transformation from one in which regulation set all prices into one in which markets for telephone equipment and long-distance service have been opened while local telephone service has (until very recently) been treated as a regulated monopoly. Experience with that sector provided the lesson that letting the regulated monopoly continue to operate in competitive markets could subvert competition in a number of ways. The regulated firm might put one firm ahead of others in the queue for getting access to the regulated service or have the customers of its regulated services bear the costs of its competitive ventures.

These concerns led in 1984 to the draconian solution of keeping most regulated local telephone companies out of the long-distance business, a restriction only slowly changing since the Telecommunications Act of 1996. In electricity, state and federal policymakers must wrestle with a similar decision: Should regulated wire monopolies be prevented from owning generation facilities? Can other operational institutions and rules ensure that transmission and distribution monopolies promote competition without forcing utilities to divest all of their generators? The widespread use of the term restructuring to describe the introduction of power competition into the electricity industry illustrates just how fundamental these concerns are.

For the foreseeable future, opening electricity markets to competition will mean that generators will be vying with one another to sell power. The transmission and distribution of that power will, however, most likely continue to be provided by regulated firms with monopolies.

As was described in Chapter 2, electricity has historically been produced and delivered primarily by investor-owned utilities (IOUs). These IOUs have operated at all levels of the production chain, generating electricity and transmitting it through proprietary high-voltage lines to their local distribution companies, which delivered the power to the final customer. A crucial question facing policymakers is whether this vertical integration remains appropriate when a utility would be selling electricity while owning the transmission lines and distribution networks that its competitors need.

The prevailing opinion implicit in the actions of electricity policymakers in the past few years is that unfettered vertical integration is not appropriate. The term for opening electricity markets to competition, *restructuring*, is itself a manifestation that expanding competition requires a change in the structure of the firms in the power industry.

An explanation of why this is so begins with a description of vertical integration and how it works in the economy at large, both when industries are fully competitive, and then when there is some concern regarding market power. Such a description identifies problems that may occur when the same firm participates as a competitor in one market (e.g., generation) while being a regulated monopolist in a related market (e.g., transmission and distribution). This is not a new problem—the massive breakup of AT&T's telephone system in 1984, and a host of judicial, regulatory, and legislative efforts before and since, were motivated by these concerns. Policymakers have a number of options (discussed below) as they deal with this problem of setting and managing the boundary between regulation and competition in the electricity industry.

Why Do Firms Look the Way They Do?

In producing any good or service, there are a number of links in the chain of production from raw materials to final sale to the consumer. Consider automobiles. One might begin with the process of making the steel and aluminum that go into cars. (One could begin even earlier and consider the process of mining the iron and bauxite that become these materials.) These metals are used to fabricate engines, bodies, frames, and other components of the car. These parts are then assembled to make cars. Trains and trucks deliver the cars to retail dealerships, which then sell cars to the final purchasers.

In this industry, and in most industries, *vertical integration* refers essentially to how many links in the chain of production a single firm undertakes. Another way one hears the idea put is as a *make–buy decision*. Do we make the component parts ourselves, or buy them in the market? Do we open our own retail stores, or sell through independent dealers? A firm that chooses to operate in multiple stages of an industry (e.g., making its own parts or operating its own outlets) is said to be vertically integrated. (For more on the economics of vertical integration, see the box on the next page.)

Economies of Scope and Scale

One of the most important concepts for understanding vertical integration is what economists call an *economy of scope*. Technically, there are economies of scope when it costs less for one firm to produce two things together than for two separate firms to produce those items separately. In electricity, for example, one would have an economy of scope between generation and transmission if it costs less for a single utility both to produce electricity and transmit it. Economies of scope are absent if generation and transmission can be provided at a lower cost when firms do only one or the other but not both.

An economy of scope can be created by something as simple as the ability to use the same factory or equipment to produce more than one product or service. For example, once one has built and staffed a hospital to provide surgical care, it may be less expensive for that hospital to provide emergency room service than it would be to build a new emergency room from scratch. However, economies of scope probably arise more often from the savings from being able to coordinate production at one stage with that at another, for example, when a car company makes both the automobile bodies and engines.

It is important to distinguish economies of scope from what might seem to be a synonymous concept, *economies of scale*. Economies of scale exist when one firm can produce a given volume of a particular good or service at a per-unit cost that is lower than if its level of production were divided among two or more firms. Economies of scale play a big role in determining how much competition a market can support. When economies of scale are minimal, many small firms can prosper; but when such economies are very large, a single firm can meet demand at least cost, potentially leading to a monopoly.

Economies of scale and scope are not the same thing. As a rule, scale economies affect how big a firm is in a market, whereas scope economies affect how many products a firm produces. Fancy restaurants may have almost no economies of scale—that is why they are typically very small. But they do have economies of scope because they are able to offer both lunch and dinner rather than specializing in one or the other. Conversely, local water delivery may have economies of scale that makes it a monopoly, but that does not imply economies of scope between water delivery and lawn care or plumbing.

The one context where economies of scale and scope may be hard to disentangle involves geography. Imagine water systems in towns A and B, which are close to each other. Conceptually, these are separate products, because someone who lives in town A can get water only from the system serving town A, and vice versa. Nevertheless, it could be less expensive for a single system to serve both towns than to have separately owned systems in A and B. Technically, this would be an economy of scope, because the products "water in town A" and "water in town B" are different. However, if the source of the economy of scope is simply that a bigger water company can deliver water at a lower per-unit cost than two separate companies, the underlying cause may be essentially an economy of scale in water production.

Vertical integration is often a matter of degree. For example, even if an automobile company does not do the actual retailing itself, it may have a long-term contract with a particular dealer to sell its cars, rather than offering them at wholesale to anyone who wants to be a retailer. The manufacturer might condition the purchase of steel on the ability to exercise some control over the way the steel is made. The automobile company might require that its dealers also offer warranty repair service, or it may require that they not sell cars from close competitors. One can think of each of these as reflecting some degree of vertical control across links in the production chain, along a spectrum ranging from zero control to full integration.

Viewed this way, almost any business is vertically integrated to some extent. Each enterprise has to decide for itself just how vertically integrated it should be.

This often requires some delicate trade-offs. The advantage of *not* vertically integrating is that one retains the flexibility to use the market to get supplies at the lowest prices from "upstream" producers, or to find the best prices to "downstream" buyers, be they other manufacturers, distributors, or retailers. This ability to be flexible and seek out the best price, however, may come at a cost. A firm may need to spend valuable time and money looking for the right option, verifying that one is getting what one pays for, and coordinating design and delivery patterns. When a firm has to build an expensive, dedicated plant to serve a particular buyer or seller, the potential for flexibility disappears.

Deciding when the gains from flexibility and good prices exceed the costs of search, verification, and coordination is a key objective of virtually any commercial enterprise. In a market economy, the general presumption is to let firms decide between using the market or vertically integrating (i.e., between buying or making), just as they can then make other decisions balancing output, product quality, and cost. If so, why treat electricity differently? Why should policymakers consider interfering with whether utilities choose to generate electricity as well as transmit and distribute it?

Monopoly Itself Need Not Be the Problem

When all of the relevant sectors in an industry are reasonably competitive, a firm's clear motive for vertical integration is to improve its ability to monitor and control costs and quality. Where there is a monopoly, however, perhaps vertical integration is not so benign.

However, the presence of a monopoly in the production chain need not make vertical integration worse. A monopolist, like a competitive firm, will perform better both from its own perspective and that of the market as a whole if it can keep its costs down. Therefore, a monopolist need not undertake this balancing between vertical integration and buying from or selling to independent firms in the market in ways that lead to higher prices or to less efficient outcomes.

A less benign way of putting it is that the amount of monopoly power in an industry depends primarily on the amount of power at any given link. If someone had a monopoly in the manufacturing of cars, for example, it need not make matters worse if it had a monopoly in selling them as well. It could make just as much money, and leave consumers just as vulnerable to high prices, by charging a very high wholesale price and letting the dealers compete among themselves over a small retail margin.

To some extent, vertical integration by monopolies could lower prices. When there are monopolies at two links in the chain, vertical integration of those links will both increase profits and reduce prices. When the separate monopoly links are not integrated, if one raises prices, the other's profits tend to shrink. Vertical integration will lead each to take the other's losses into account, and thus keep them from raising prices as much as they would have acting separately.

Economists and antitrust experts differ about how much vertical integration matters beyond these basic ideas. In some cases, vertical integration may create a monopoly at one link where there was not one before, for example, if a car manufacturer signed up every dealer in a metropolitan area to carry only its cars and none of any competitors. A more contentious speculation is that vertical integration may change the strategic options open to firms with market power in ways that make it harder for others to compete with them. For example, vertical integration that enables a monopolist in one market to enter another market may eliminate a pathway that others might use to eventually compete with it. Such a claim underlies the allegation of Microsoft's opponents that it tried to take over the Internet browser market to prevent future competition in computer operating systems.

The Regulatory Wrinkle

In the electricity market, the concerns raised by vertical integration are less speculative. Regulation of the "wire" monopolies in the electricity industry to limit their market power, perhaps paradoxically, creates anticompetitive incentives for vertical integration.

In essence, vertical integration becomes potentially more troublesome when a firm has a regulated monopoly in one market and operates on an unregulated basis in another. Crossing the regulated–unregulated boundary may enable the monopolist to evade the regulation and exercise the market power the regulation was instituted to control. The story is not new. Concern regarding these incentives motivated the 1984 corporate breakup of AT&T's telephone operations, along with much of the regulation of that industry before and since. The story is not even new to electricity; these same concerns partly motivated the 1935 Public Utility Holding Company Act (PUHCA).

Early concerns in electricity had to do with the possibility of excessive *transfer pricing*, that is, the prices at which a vertically integrated firm sells supplies or service to its regulated businesses. The costs associated with the transfer prices are used to justify rate increases; the revenues show up as profits on the firm's unregulated operations. PUHCA set limits on the interstate structure of utilities in part to prevent overcharging central administrative services to the regulated utilities in particular states. Another example, perhaps more hypothetical than real, was that a utility might integrate into coal or natural gas and sell the fuel to itself at above-market prices, raising electricity prices.

A variation on this theme, known as *cross-subsidization*, became prominent in looking at telecommunications. Cross-subsidization refers to the practice of charging expenses incurred to provide an unregulated product to the books of the regulated service. Cross-subsidization can lead to higher rates for the regulated service. It can also displace competitors in the unregulated sector that have lower "real" costs than the regulated firm but that cannot compete against the subsidies. An example in the electricity industry was the case *Cantor v. Detroit Edison* (428 U.S. 579, 1976), in which the Supreme Court found that Edison's giveaway of lightbulbs was anticompetitive and, incidentally, not immune from lawsuit simply because Edison was regulated by the State of Michigan.

The concern receiving the most attention, however, is probably that the owner of the regulated monopoly would discriminate in favor of its own unregulated operations vis-à-vis its competitors. The now classic example of this is in the telephone industry. The government forced AT&T to spin off its local telephone companies, and then kept those divested companies from reentering the long-distance business, out of a concern that unaffiliated long-distance competitors and their customers would get slow or noisy local telephone service. In electricity, the concern would be that a transmission or distribution grid owner that also generates power would be slow to provide access to, maintain, or eliminate congestion on lines its generation competitors need.

In all of these stories, both monopoly and regulation are key components. If the firm could charge whatever it wanted for its monopoly service, it would have no need to discriminate to get profits indirectly by favoring its unregulated affiliate. Similarly, it would not have any reason to overcharge for supplies or cross-subsidize other operations just to raise its regulated rates.

Regulate, Restructure, or Divest?

A regulator faced with trying to promote competition in the face of vertical integration across the boundary between regulated and unregulated sectors has three main options; in order of increasing severity, these are to monitor and regulate dealings between the firm's regulated and unregulated operations, to restructure, or to divest. Each option has its advantages and disadvantages. Generally, the higher the fence between the regulated and unregulated sectors, the easier it is for the regulator to protect competition and consumers, but the harder it is for the firm to take advantage of any legitimate cost savings or operational improvements associated with vertical integration.

That a regulated transmission or distribution company that also generates and sells electricity wants to cross-subsidize, discriminate, or set high transfer prices does not mean that it can do so, if the regulator is able to pay sufficient attention to its actions. High transfer pricing works only if the regulator cannot tell that the firm is paying too much for the services it is purchasing, and it pays only if the regulated firm purchases a lot of its supplies from the unregulated sector. Here, the transmission company would have to use (not just ship) a lot of electricity, and pay an above-market price for it without being detected by the regulator. At first blush, neither is very likely. Electricity markets appear to be evolving in ways in which the market prices are quite public, and a regulator should be able to tell fairly easily if the transmission or distribution side is paying too much for it. Also, the quantity of electricity used in the regulated sectors would seem to be quite small. However, ancillary services (described in Chapter 2) may be an important exception. We consider those markets in more detail in Chapter 10.

Cross-subsidization requires a regulator to apply cost-of-service regulation so diligently that increases in reported cost readily lead to increases in rates, yet also to be unwilling or unable to determine whether reported expenses were actually incurred to provide the regulated service. Audits necessary to detect cross-subsidization will require more effort, to the extent that the labor or supplies to provide generation that are used to inflate regulated rates are of the kinds that are used to provide transmission or distribution. Cross-subsidizing purchases of fuels or generator turbines would be relatively easy to detect and prevent. An ever-present concern is that a transmission or distribution utility could use its secure standing in capital markets to take on market risks associated with competitive "merchant" generation, passing on the cost of the higher rate of return associated with that risk in the form of higher prices for its regulated services. But as is discussed below, the greatest potential for cross-subsidization may involve businesses outside the electricity sector.

In addition, the method of regulation can make a difference regarding how much one needs to worry about transfer pricing or cross-subsidization. Those tactics allow one to evade regulatory limits on monopoly prices and profits by taking advantage of a linkage between (reported) costs and rates. If the regulator adopts "incentive regulation" (discussed in Chapter 8), prices depend less on costs, thus weakening this link. There is no reason to charge oneself too much for services or supplies, or to shift costs from the unregulated sector to the regulated one, if costs make no difference in one's rates.

Discrimination is probably the most substantial concern justifying stronger boundaries between the wires and generation sectors. Most of the attention, particularly at the federal level, has involved transmission. Regulators may find it difficult to determine whether a transmission utility's decisions regarding line maintenance and installation were legitimate or strategically chosen to hurt the reliability of competing generation services. The transmission utility could always say that its generation is more reliable and cost less because of better management, not discriminatory access to the grid. Incentive regulation in and of itself does nothing to discourage discrimination, although the higher the regulated price, the less the firm will find it profitable to stand in the way of competitors that rely on and purchase it. (For more on regulated firms, see the box.)

Structural Remedies in Transmission

Corporate "Firewalls"—the Independent System Operator

If more attentive or incentive-based regulation does not suffice, the next step is to turn to

Buying Market Power from a Regulated Firm

Separating generation from the regulated transmission monopolies has been an important policy tool to protect competition in electricity markets. The central idea is to make sure that those who own generation are not able to control the transmission grid so as to favor their own operations. With such control, the generation company could effectively use the transmission company's monopoly power to create a monopoly in electricity, leading to just the outcome that regulation of the transmission system is designed to protect.

The usual context in which one might be concerned that a generation company could acquire control over a transmission facility is if they are in the same vertically integrated company. This is the underlying rationale for either formal separation of generation and transmission by having transmission provided by a stand-alone transmission company, or through functional separation by having transmission service provided through an independent system operator (as is discussed in the text).

Unrestricted vertical integration need not be the only way a generator might acquire such control. Generators need to obtain the rights from transmission companies to send electricity through their lines. Through long-term contracts, a single generation company could obtain control over enough of the capacity of a transmission system to limit the ability of its competitors to send their own electricity through the grid. This control would give this generator the ability to raise prices without being constrained by competitive pressure from other generators. Competition could be similarly challenged if an electricity retailing company acquired sufficient grid capacity to limit its competitors' ability to obtain energy to sell to residential, industrial, and commercial customers.

Concerns similar to these have been voiced at the Federal Energy Regulatory Commission regarding the possibility that a firm competing in the sale of natural gas acquired control over sufficient capacity of regulated gas pipelines to create the potential to raise gas prices above competitive levels. Along with implementing structural policies to separate regulated monopoly facilities from related competitive markets, regulators may need to monitor contracting between competitors and the regulated firms (e.g., generators and transmission grids) to ensure that no competitor acquires effective control over the regulated facilities.

structural solutions. The less drastic step is to allow the firm to retain interests in both transmission and generation, but to impose assorted restrictions on the control associated with that ownership. Before the breakup of AT&T, telephone regulators were in the process of implementing regulations to impose a formal separation between competitive and regulated sectors. The term of art in that context was the *fully separated subsidiary*. The thought was that creating separate companies within the company, with separate employees and books, would make it easier for the regulator to monitor and prevent discrimination, cross-subsidization, and inappropriate transfer pricing.

In electricity, a more stringent form of separating operations within an integrated company is to vest management of the transmission grids with an *independent system operator* (ISO). So far, several ISOs have been formed. Most of the ISOs formed before 2000 were those that corresponded with major "power pools." These power pools, essentially joint ventures of the utilities serving a particular region—such as New England, New York, or the "PJM" pool serving Pennsylvania, New Jersey, and Maryland—already served to coordinate power exchanges among the utilities to promote reliability. In response to Federal Energy Regulatory Commission (FERC) Order 2000 and subsequent policy statements promoting the creation of larger *regional transmission organizations* (RTOs; see box on next page), several other ISOs and RTOs have been proposed for FERC approval.

Whether an ISO is built on existing power pools or started from scratch, it is a nonprofit operational entity typically directed by a board comprising incumbent utilities, other independent power generators, and other interested consumer groups and public agencies. The ISO's structure is intended to promote decisions that do not favor the incumbent utilities or impose unnecessary costs on generators that want to enter the market. Although ISO rules have to be approved by FERC (see box), there remains sufficient variety in approaches to see which if any formats ensure appropriate independence of the transmission and distribution sectors. However, the complexity and controversy surrounding ISOs so far has little directly to do with protecting against discrimination. In addition to ensuring nondiscrimination and preventing cross-subsidization, ISOs across the country are frequently charged with management of the markets in which generated power is purchased and, in many cases, dispatched. The potential reasons for that expansive role are considered in Chapter 10.

Divestiture—the Transco

The more radical procedure for ensuring independence is simply to prohibit vertical integration between generation and transmission or distribution. This would involve forcing the vertically integrated utilities to sell off either their generators or grids. Either way, one would be left with a stand-alone entity—the *transmission company* (transco)—that would be for-profit but regulated. As of the summer of 2001, three separate for-profit transcos had been proposed to FERC by utilities in Florida, utilities in North and South Carolina, and the Southern Company (which serves customers in Alabama and Georgia) to take on the function of an RTO in those regions.

The transco option may be more effective in preventing discrimination and the other anticompetitive practices that might occur with vertical integration across

Regional Transmission Organizations

The Federal Energy Regulatory Commission (FERC) has been encouraging formal separation of the control of transmission assets from generation assets ever since it issued Order 888 in 1996. In this order, FERC initially endorsed but did not mandate the creation of independent system operators (ISOs), establishing principles that should govern the conduct of an ISO before FERC would authorize it to operate transmission facilities under its jurisdiction.

Believing that a stronger role for ISOs was necessary, and that they could cover larger geographic areas, FERC went further in Order 2000, issued in 1999. In that order, FERC endorsed the principle that all transmission facilities should be placed in the hands of regional transmission organizations (RTOs), either ISOs or transcos. FERC stated that such RTOs should:

- be financially and operationally independent of market participants;
- support a region of sufficient scope to be effective and to support efficient, nondiscriminatory electricity markets;
- monitor, operate, and maintain security of transmission facilities under its control; and
- maintain short-term reliability, including the right to order redispatch of generation and to approve scheduled outages of transmission facilities.

FERC also listed eight minimum functions that an RTO should be capable of undertaking:

- have sole authority to provide transmission service, administer its tariffs, and approve interconnection;
- create market mechanisms to manage transmission congestion, including tradable rights for transmission access and opportunities to hedge against the risk of high congestion costs;
- develop and implement procedures to deal with loop flow (discussed in Chapter 8);
- be a supplier of last resort of ancillary services (defined in Chapter 2);
- provide information on transmission system conditions, capacity, and constraints to all market participants in a timely manner;
- monitor wholesale electric power markets to identify problems with market design and market power;
- alleviate congestion and promote expansion by encouraging market-based operation and investment, accommodating multi-state agreements, and meet FERC's planning requirements; and
- coordinate activities with neighboring regions, whether or not their transmission systems are managed by RTOs.

In Order 2000, as in Order 888, FERC again stopped short of requiring that utilities create or join an RTO, allowing them to file descriptions of obstacles to RTO participation in lieu of proposals to form or participate in RTOs. However, as noted in the text, FERC subsequently has stated its intention to encourage the formation of a few large RTOs that together would cover the U.S.

the boundary between regulation and competition. However, it is more extreme and may be more difficult to bring about. Whether it is necessary to go beyond the ISO to a transco depends partly on whether the ISO's rules and regulatory oversight will adequately prevent discrimination and promote market efficiency. Some observers have expressed a concern that the nonprofit nature of ISOs may make them more susceptible to political pressure that could prevent providing transmission and distribution service at least cost.

Whether an RTO takes the form of an ISO or transco, an important policy question is its appropriate geographic breadth. Because the entire transmission grid serves as a single economic unit (see Chapters 2 and 6), the ideal RTO perhaps should cover an entire intertie, and not a subset of a transmission grid. As a step in that direction, FERC in July 2001 announced its intention to have northeastern, southeastern, midwestern, and western areas of the United States each served by a single RTO.

Voluntary Divestiture

One possibility to watch is that utilities themselves may solve the problem by voluntarily divesting generation, at least in those regions where they own transmission and distribution facilities. Several utilities (including Potomac Electric Power Company, operating in the District of Columbia and Maryland, and General Public Utilities of Pennsylvania) have made such voluntary divestitures and chosen to focus on the regulated portions of the electricity supply business. As long as a utility remains vertically integrated, it risks being subject to regulatory oversight that could inhibit exploiting new business opportunities in their industries. Freedom from such oversight was partly why AT&T elected to settle the 1980s antitrust case and divest outright its regulated local telephone monopolies.

One of the options these utilities may have in mind is to expand their wire-oriented operations to enter telecommunications markets, such as local telephone service, cable television, and high-speed Internet access. The federal government encouraged such efforts with the 1996 Telecommunications Act, which (among many other things) allowed utilities to create "exempt telecommunications companies" free of some of the restrictions PUHCA placed on utility expansion into other markets.

However, state and federal regulators may need to keep a watchful eye on these efforts because of the possibility of cross-subsidization. Utilities may be tempted to charge the costs of entering these telecommunications markets to their electricity customers. One hypothetical example would be to install a high-speed communication network ostensibly to monitor electricity use on a location-by-location basis, charge the full cost of those lines against the rates charged for distributing electricity, and then use the lines to provide other communications services. But one needs to be careful not to discourage capital investments that may generate legitimate economies of scope and increase competition in cable television and local telephone markets that have previously been monopolies themselves.

CHAPTER

Regulating Rates for Transmission and Distribution

The wire segments of the electricity sector—transmission and distribution—will continue to be regulated (as we saw in Chapter 6). Regulation, of course, is not a new issue in this industry, but most regulation before restructuring has been devoted to setting the electricity rates that users pay. With restructuring, power prices will be set by the market, with prices users pay directly or indirectly including those power prices plus the regulated charges for delivering electricity from the generator to the customers' premises. This chapter first discusses methods for setting rates for transmission and distribution, describing both traditional "rate-of-return" regulation and new "incentive-based" methods that could lead to lower costs and more efficient operation.

Although these principles apply to both distribution and transmission, the latter presents difficult problems. A generator may have to go through lines owned by several utilities in various states. Transmission prices might best be set by broad geographical regions and independent of distance, or perhaps should include charges that increase with distance or the number of times the path crosses a state line or uses a different utility's facilities. An even more complex set of questions related to pricing and incentives is associated with the possibility that transmission lines may be congested.

Local distribution and long-distance transmission of electricity are likely to be regulated for the foreseeable future (as was noted above). If so, regulators will need to figure out how to set the prices generators and users pay to get power delivered. Because regulators had set electricity prices before the opening of wholesale and retail markets, this is not an unfamiliar question. State public utility commissions essentially added up all the costs of generating, transmitting, and distributing electricity during peak and off-peak times, calculated an average cost per kilowatt-hour produced, and set a price. In addition, the Federal Energy Regulatory Commission has for some time been setting the rates utilities charge to transmit electricity sold in wholesale power markets.

With prices for electricity generation set through competition, restructuring affords an opportunity to rethink how to devise separate rates for transmission and distribution (T&D). Questions regulators need to address include:

- the *level* of rates—how much revenue should be raised overall to pay for T&D;
- the *adjustment process*—how rates should be adjusted over time in response to actual or expected changes in the cost and profitability of T&D; and
- the *structure* of rates—how the burden of covering this revenue should be based on dimensions of time, distance, congestion, the location of the generators and customers, the amount of power transmitted, and the type of customer (e.g., industrial, commercial, or residential).

Trade-offs are present throughout. Do we try to keep rates as low as possible, or do we give regulated firms the opportunity to make more money as an incentive to cut costs and innovate? Should charges depend on how much power consumers use, or should we raise money through fixed, up-front fees that are the same regardless of how much power is used? Should we set high prices to discourage generators from using congested lines, or would this provide an incentive to keep lines congested?

We start by looking at traditional rate regulation, alternatives that may give T&D companies more incentives to control costs, and questions regarding how to structure rates. We conclude by turning to a couple of issues that are paramount in the transmission context—whether to set prices based on the location of generators, particularly when lines are congested.

Setting the Level: Traditional Rate Regulation

The typical method for setting rates in the electricity industry (and other regulated industries) during the past century has been to employ a three-step process:

- Estimate the amount of the regulated service one expects to supply, usually over the course of a year.
- Estimate the cost of supplying that amount of service during the year.
- Divide the latter by the former to get the price of electricity.

Because of the second step, this method of regulating goes by the name of *cost-of-service regulation.* In practice, that second step itself is quite complex. To calculate the cost of service, a regulator must determine:

- *operating expenses,* such as labor, fuel (in the case of regulated electricity generation), and other consumables that get used up in the course of a year;
- the *rate base*—that is, the level of investment in capital equipment, from transmission towers and lines to vehicles and furniture, that lasts for a number of years;
- the amount of *depreciation*—that is, the amount of the investment in plant that investors should get back in any given year (this may or may not be related to any decline in the economic value of the plant);

- the allowed *rate of return* investors can earn on the *undepreciated* rate base—that is, the investment that has not already been returned to them (depreciated) in earlier years; and
- corporate income taxes payable on the income to the investors.

In the electricity sector, other expenses have been included in the cost of service. Examples include funds to subsidize residential purchase of high-efficiency appliances (demand-side management programs; see Chapter 16) or to cover future costs of disposing of nuclear fuels.

In regulatory proceedings, the most controversial part of the process is the determination of the allowed rate of return. Other costs are either directly observable in principle (e.g., operating expenses) or are prescribed via accounting rules (depreciation rates) or tax laws. However, the rate of return generally should reflect the cost of capital throughout the economy, forecasts of inflation, and the premium that should be added to reflect any risk that investors may not recover their full investment. The controversy associated with all of these questions has led this method for setting price to be known as *rate-of-return regulation*.

Issues with Rate-of-Return Regulation

For almost 40 years, economists have been writing about some of the problems associated with rate-of-return regulation. To get it right, the regulator needs to get data on demand and costs. Regulation, to be effective, requires that the industry be relatively static. If costs and demand are changing rapidly, the regulator may not be able to gather these data and estimate prices accurately enough to ensure that consumers do not pay excessive prices and that the regulated firms receive enough revenue to cover their costs.

In addition, the regulated firm, the main source of the information, has every reason to fudge for its own benefit. In a relatively static industry, regulators are better able to audit the firm's operations and verify that expenses are accurately reported. They also may be able to compare the data they compute with that computed for similar firms in other states, as a check to prevent misrepresentation of costs.

Another difficulty with rate-of-return regulation is that the regulator typically has to forecast demand to calculate costs, and then to prescribe a price. But, typically, one does not know how much of the service buyers will want until we know the price. Regulators would then need not only to be able to add up costs, but have the information and ability necessary to come up with statistical estimates of how its pricing decision might affect demand.

The desirability of avoiding the circularity of using demand to find a price and then using price to predict demand is why regulation works best when the product is one that people tend to purchase in roughly the same amount regardless of price. If demand is relatively independent of price—what economists call *inelastic*—the regulator is better able to predict demand with reasonable accuracy than if demand is sensitive to price. If consumers will take or leave the product depending on its price, the complex "demand determines price, which determines demand, which determines . . . " circularity is unavoidable.

The most studied difficulties with rate-of-return regulation, however, involve the ability of the regulator to specify correctly the appropriate rate of return. The goal would seem to be to give the investors just enough to compensate them for putting up their money in the regulated enterprise. Giving them too little would discourage them from making the investments for services they need, whereas a high allowed return would lead to prices for the service being greater than they need to be to provide appropriate compensation. In addition, if the allowed rate of return exceeds the costs of capital, the firm has an incentive to increase profits by installing too much plant (i.e., "padding the rate base" or "gold plating" the plant).

Transmission and distribution do not appear to pose particular difficulties in verifying operating expenses, applying depreciation rates, understanding the relevant tax code, and estimating an appropriate return. To the extent that transmission and distribution incorporate relatively stable technologies and costs, these difficulties may be less severe than in other regulated industries, such as local telephone service. Nonetheless, rate-of-return regulation is subject to some inherent flaws, perhaps paradoxically, when it works perfectly as designed.

Addressing Inefficiencies through Incentive Regulation

Suppose that costs and demand are estimated correctly, and that prices are set so that revenues just cover expenses. If revenues always just cover expenses, the regulated firm has no incentive to be efficient. If it fails to control costs or to undertake worthwhile innovations, it loses nothing, because it can still charge prices high enough to prevent any losses. If prices fall after cost reductions or innovations, the incentive to be efficient falls as well. The potential for waste has forced price regulators to get involved in the planning and approval of regulatory investments. Regulation becomes not the relatively simple matter of calculating and enforcing a price, but forces the regulator to take an active role in the management of the regulated firm itself.

To encourage innovation and cost cutting, and to extricate regulators from the management of the firms they regulate, economists have devised and implemented an alternative form of regulation known as *incentive regulation, incentive-based regulation*, or *performance-based regulation* (see box on next page). The defining feature of incentive or incentive-based regulation is that prices remain regulated but are divorced from actual costs. It seems like a paradox that regulation could lead to lower prices if those prices are set without regard to cost. However, doing so allows firms to keep any cost savings they can achieve.

The most common form of incentive regulation in the monopoly context, employed primarily in telecommunications, is known as "price cap" regulation. With price caps, the regulator sets an initial price and then specifies in advance how that price will change during a period of years to allow for inflation (which would increase prices) and expected productivity (which would reduce costs). The regulated firm then is free to decide how best to cut costs and innovate to make money, taking these prescribed prices as given. Because prices are fixed in advance, rather than falling if costs fall, any dollar the regulated firm saves, it keeps. This restores the incentive to be efficient that conventional cost-of-service regulation impedes.

Incentive Regulation and Competition

Incentive regulation may seem less paradoxical if we see that in important ways it mimics competition. The defining characteristic of a competitive market is that—from the perspective of any individual competitor—the price is set by the "market" rather than any particular firm. If a firm tries to charge more than the market price, its customers will run to its competitors. And because the customers are already paying the market price, the firm can pretty much sell at that price as well. It makes money as long as the cost of producing more goods or services for sale—what economists call the *marginal cost*, as distinct from *fixed costs* that do not change as production changes—is less than the market price.

How closely the electricity industry, or any other industry, matches this ideal portrait of competition, is frequently quite controversial. Chapters 7 and 9 present some of the regulatory and antitrust policies that could help bring the industry closer to this ideal.

When prices are set by competition, rather than by the unilateral decision of a seller insulated from the checks and balances of the market, the firm is what economists call a *price taker*. Among the important consequences of being a price taker is that if a firm cuts costs, it can still sell at the same market price. For example, suppose competition among pizzerias sets the price of a pizza at $10. If an individual pizzeria can cut its costs, it can still sell pizzas at the prevailing $10 price. If spending $200 on a new piece of equipment can cut its overall costs by $300, it gets to keep the $100 net gain. It therefore has the incentive to make the effort and investments necessary to cut costs by a greater amount—just what we would want.

Incentive-based regulation tries to recreate these same conditions in a regulated industry. Suppose the regulator can set a price without regard to the costs the regulated firm actually incurs. The regulated firm is now a price taker itself, and thus has the same incentives to be efficient and innovate as would firms in a competitive market.

How does this help consumers? In a competitive market, like that for pizza, as more and more pizzerias adopt cost-cutting practices, the market price will fall as competition among them drives down the price. (Note that firms are still price takers, in that competition would drive down the price even if an individual firm chose not to cut its costs.) Getting the benefits to consumers in a regulated industry is trickier. Ideally, the regulator sets prices to fall over time in a way that reflects how much the regulated firm *would be expected to* cut costs. But those prices do not depend on how much the firm actually *does* cut costs, to ensure that the firm keeps each dollar of costs that it saves.

Is the Cure Effective?

In its Order 2000, the Federal Energy Regulatory Commission (FERC) encouraged but did not require regional transmission organizations (see Chapter 7) to adopt incentive regulation (performance-based regulation) for transmission rates. In not making performance-based regulation mandatory, FERC recognized that incentive regulation is not a panacea. In separating prices from costs, the regulator creates the possibility that prices may not be as low as they could be, and still have the firm recover its costs. These higher prices may discourage some customers from using the regulated service, when it would be cost-effective for them to do so. In addition, higher prices typically mean higher profits for the firm at the expense of the customers, which may go against prevailing beliefs about who should benefit from having regulation in place.

Many are also concerned that incentive regulation may affect the quality of service. The view is that if firms have an incentive to cut costs, they will have an incentive to reduce the quality of service they offer. Sometimes price controls do lead to reductions in quality, but primarily when the price is below the level that would be set in a competitive market. In those cases, sellers lose money by continuing to sell, and may find that reducing quality is the only way to cut demand and stay in business profitably.

In regulated industries such as electricity distribution and transmission, however, prices are typically above the cost of providing service to a particular group of customers for a particular time, because revenues must cover both those marginal costs of service and the fixed cost of the entire grid as a whole. For that reason, if a grid were to cut quality, say, by making the system more prone to outages (see Chapter 11), the money it would lose because of the forgone opportunity to sell its services at the regulated price while the lines were down would be greater than any money it would save from not having to provide service. Accordingly, incentive regulation need not dampen the incentive to provide reliable service.

The major difficulty with incentive regulation is operational. For it to work, the regulated firm must believe that if it cuts costs, the regulator will not renege on the deal and force it to cut prices. Similarly, it must believe that if it fails to meet expected productivity targets and thus loses money, it cannot go to the regulator and get a price increase. It may be too much to expect a regulated firm to have these beliefs. If it is too successful at cutting its costs, its customers will put political pressure on the regulator to cut its prices. If it is unsuccessful at controlling costs, its regulator may be faced with the unpalatable choice to let it go out of business or to let it increase prices. Either way, the commitment that prices will not follow costs, and the rewards from cutting costs, are weakened.

In practice, price cap regulation involves revaluation of prices after a few years, so the commitment to separate prices from costs and let the regulated firm sink, swim, or soar above the water remains truncated. Recognizing these limits, regulators have adopted hybrid profit-sharing methods that preserve some of the benefits of incentive regulation, while recognizing political limits on how much money a regulated monopolist can make. Under a scheme widely adopted in the telecommunications industry, the regulated firm may be allowed to keep profits below a certain rate of return, but if it goes beyond that, it must give back some percentage of the profits in the form of lower prices.

The key factor in deciding whether to adopt incentive regulation as opposed to cost-of-service methods is the degree to which the need to respond to changing technology, demand, and competitive conditions is important. If these are relatively stable, rate-of-return regulation may be adequate. In more dynamic settings—where the regulated firm should investigate new technologies, make use of new ways to control costs, and expand in response to changing market conditions—incentive regulation is more likely to be beneficial. Even a small cost savings can swamp any adverse effects from prices being a little higher than the ideal. In addition, if the market is likely to become competitive in the reasonably near future, price cap regulation can be a good transitional device, by creating incentives for efficiency while minimizing the need for government oversight. Proposals to apply temporary price caps to wholesale electricity markets, following the California crisis, invoke some of these issues; we discuss such caps in more detail in Chapter 9.

For these reasons, many state and federal regulators have been adopting price caps and hybrid methods to regulate telephone companies. Telephone service is becoming more and more technologically open, and optimism remains high that all sectors of that industry will become competitive in the near future. Whether the prospects for technological change and competition in the transmission and distribution of electricity make such methods useful for regulating those industries remains a matter for regulators to judge.

Structuring Prices

When electricity was regulated end to end, electricity regulation was typically designed to come up with a simple price per kilowatt-hour of energy used. But when we regulate only a part of the industry—the transmission and distribution grids—a variety of possible methods for setting prices presents itself. Do we base fees on how much power people use? Do we charge flat fees independent of actual usage (so-called license plate fees)? How should the distance between generators and users figure into the fees to use a grid? Should we charge all power users the same price, or should some pay more than others?

To explain the general advantages and disadvantages of the different methods of structuring prices, we first look at the relatively simple case in which transmission and distribution grids are able to transmit all the electricity that generators want to send to their customers. (We turn below to the considerably more complicated matter of how to set rates when some of the lines, particularly in the transmission grid, are congested.) When there is no congestion in a grid, the cost of sending one more kilowatt into a transmission or distribution grid—the marginal cost of using the grid—is negligible.

If marginal costs are negligible, the ideal price for using the grid should also be zero. Forcing users to pay a positive amount would unduly discourage them from using the grid. To illustrate this, imagine that the fee for distributing electricity was $10 per megawatt-hour (MWh). This price would typically be added on to the cost of the power itself. Someone for whom the value of getting electricity was only $5 per MWh above the cost of the power would find it not worthwhile to pay the $10 charge. However, the $5 benefits of distributing that MWh of electricity to him exceed the negligible cost of distribution, and from an economic standpoint, it would be inefficient not to do so.

If transmission and distribution are free of charge, however, grid owners and operators will get no revenue. They will not be able to cover the considerable fixed costs associated with making their grids available. To cover these fixed costs, each generator and customer could pay a *fixed fee*, on an annual or monthly basis, independent of the amount of power they use. Once they pay this fee to get on the network, they would pay no more for the transmission and distribution of power, and thus would use it efficiently.

The problem, of course, is in setting these fixed fees. Should big generators and big customers pay larger fees than small generators and small customers—for example, by paying a fixed fee for the right to transmit up to some fixed amount of power through the grid? This sounds fair, but then one is basing the fee to some extent on usage, at least the maximum amount of transmission capacity that an electricity supplier or customer expects to use. An equal fee paid by all, however, will unduly discourage the construction of small generators or the survival of

Basing Fees on Willingness to Buy: Ramsey Pricing

To the extent that a firm charges usage-based prices above (zero) marginal cost to cover its fixed costs, a method favored by economists over the years has been to set prices higher for customers who are less likely to cut back on their use as the service becomes more expensive. The origin for this idea came from a British economist and philosopher, Frank Ramsey, who investigated how one might institute a tax system to minimize the costs to an economy of raising a predetermined amount of revenue. It is the same problem—only on a smaller scale—to decide how to charge different groups of customers varying prices (or how to charge varying prices at different times of the day or year) for a regulated service, so as to cover fixed costs while minimizing losses from discouraging use of that service. The fees that minimize these losses, accordingly, are frequently referred to as *Ramsey prices*.

The idea behind Ramsey pricing is that the economic harm from setting prices above cost is in the reduced purchases such high prices discourage. Applying this principle, one should charge higher transmission or distribution fees to those willing to pay more for electricity and who would continue to purchase power even if those fees were to rise. One could also imagine fees charged to generators, where low-cost generators would pay more because fees are less likely to lead them to reduce electricity production. Similarly, one would not want to charge high distribution fees to, say, a large firm that might be in a position to relocate its operations if the cost of getting electricity were to become too high.

These applications illustrate only some of the problems with this sort of demand-based pricing. It may be politically difficult or impossible to charge different groups different prices for the same thing, although it does happen. Industrial users of electricity have been getting substantial discounts, in no small measure because regulators will be sensitive to both their direct clout and their threats to relocate. Charging higher fees to low-cost generators may be fine in the short run, but having to pay higher transmission fees would obviously discourage power companies from cutting their costs.

smaller businesses. Such a fee would create an artificial economy of scale that gives an advantage to bigger power users, which can spread the fee over larger levels of production. (For a view of how willingness to pay might be factored into fees, see the box.)

A possible compromise is to raise some, but not all, of the cost of providing transmission through fixed fees and the rest through prices paid for each kilowatt-hour generated or used. These are called *two-part tariffs*, reflecting both a fixed component and a per-unit part of the fee. In principle, one can always achieve more efficient use of a distribution or transmission grid with two-part tariffs than with exclusively usage-based prices. A harder question, though, is whether these benefits are worth the administrative costs of figuring out the right set of fixed fees and the political costs of imposing them.

Price and Distance in the Transmission Grid

A defining characteristic of transmission grids is loop flow, the fact that power inserted into a transmission grid will take all paths to get to its destination (see Chapter 2). Because electricity takes all paths to get between two points, the boundaries between the wires owned by company A and those owned by company B, or those in state S and those in state B, are effectively meaningless. The price of transmission therefore should, again ideally, not vary depending upon whether the pathways between the generator and the user cross corporate or political boundaries.

This suggests that prices for transmitting electricity throughout a grid should be akin to the way we set prices for delivering mail in the United States—it costs customers the same to send a letter from Manhattan to Alaska as it does to send one to Brooklyn. It is not surprising, then, that this form of pricing transmission is known as *postage-stamp pricing*. To a first approximation, the price of transmitting power between a generator in one area and a user in another should not be determined by *pancaking* (i.e., adding up the separate transmission rates charged by every transmission company owning facilities between the two). FERC's Order 2000 prohibits pancaking within areas controlled by regional transmission organizations (RTOs). Pancaking could still occur for power transmitted across more than one RTO, but FERC is encouraging RTOs to adopt reciprocal waivers of transmission access charges. In addition, not every transmission utility is yet required to join an RTO.

Complicating the matter, however, is that some electricity is lost in the process of transmission, and more is lost the farther it goes. Although electricity takes all paths between two points, it will use the segments of least electrical resistance more than those of higher resistance. As it uses these lines, some small percentage of the power is inevitably lost in the form of heat in proportion to this resistance as it goes through the lines. Roughly speaking, resistance on any segment is proportional to its length. The resistance between any two points therefore increases with the distance between them, although it falls with the number of different alternative routes available, all else remaining equal.

Resistance and line losses imply that a generator is more likely to use transmission facilities near it than far away. How quickly the power drops off is a complicated and specific technical question. But the losses suggest that, to some extent, the average cost of the lines used by a particular generator will tend to be based on those relatively near to it rather than on those in some distant part of the country. Viewed in the long run, the added transmission capacity built to serve an additional generator or user will be more likely to be close to that generator rather than farther away. Accordingly, transmission rates could be regionally based to recover costs from charges to the generators that are more likely to use relatively nearby lines. Rates may not meet the ideal of being postage-stamp based and free of pancaking.

The choice between pancaking or postage-stamp fees is present even if fees are independent of usage. Fixed fees could be the same for every generator using a transmission grid covering a wide area. Generators could also pay separate pancaked fixed fees for rights to use capacity on transmission grids owned by separate utilities. In between, one could have fixed fees to pay for rights to use an entire regional transmission system, but those fees would vary by the location of a generator—just as one can get a driver's license in any state that allows one to drive everywhere in the United States, but the fees vary from state to state.

Transmission Congestion

The above observations on pricing assumed that the transmission grid was large enough everywhere to carry all of the electricity between the generators and their customers. However, that need not be the case. If transmission lines carry power beyond their designed capacity, they can overheat and fail.

Because of loop flow, the possibility of congestion means that holding line usage within capacity constraints leads to peculiar, counterintuitive relationships between demand for power and dispatch of generating capacity. Without getting into the technicalities, here is an example of the sort of thing that can happen. Suppose that we have generators at points A and B. The generators at A can provide energy at $20 per MWh, and the generators at B can provide power at $30 per MWh. Suppose also that we have users at location C, who demand more electricity at $20 per MWh than the generators at A can provide. To get the generators at B to produce enough to meet the needs at C, the price of power they receive must be at least $30 per MWh.

If the lines are not congested, the price of electricity at C should be (ignoring line loss) $30 per MWh. This is because the cost of making 1 more megawatt (MW) of power available at C just involves having the generators at B supply 1 more MW, and the cost of doing so is $30 per MWh.

However, suppose that the transmission lines between locations A and C are at their capacity. Suppose further that the transmission grid includes lines that run between A and B. Then, if the users at C demand one more unit of energy, we cannot simply have the generators at B increase their supply. If they do, loop flow implies that some of that power will try to go over the congested line between A and C. Were that to happen, the A–to–C line would fail. To get that additional kilowatt to C, then, the generator at A has to cut its supply, and the generator at B has to increase its supply to make up for A's reduction. They need do this to keep the total amount of power going through the A–to–C line at no more than its capacity.

The size of this adjustment depends on the resistance of the various segments in the transmission grid. The relationships that follow are complex and in some cases counterintuitive. It is not hard to come up with examples in which getting 1 more MW of power to C would require that A *cut* its production of power by 1 MW, and B increase its supply by 2 MW. If so, the cost of getting one more MWh of energy to C will be twice $30 per hour, the cost of generating 2 more MW of power at B, less the $20 per hour saved by reducing A's power production by 1 MW. Toting this up gives a marginal power cost of $40 per MWh—more than the cost at A or B!

Nodal Pricing

Getting generators to produce power so as to get it delivered where users want it without overtaxing a congested transmission grid requires a set of transmission prices that are specific to each location. The price of power at C should be $40 per MWh, because that is the real cost of making it available to them. One way to do this is to charge a congestion fee of at least $20 per MWh to the generators at A, so that they do not try to expand and sell more power where people are willing to pay $40 per MWh for it. If so, the generators at B should see a transmission price of at least $10 per MWh when the A–to–C line is congested.

Alternatively, one could charge the users at C $10 per MWh used for transmission, which would set the power market price at $40 – $10 or $30 per MWh. With a $30 price, the generators at B will produce the right amount of electricity if they see no additional congestion fee, and those at A will produce the right amount if they face a $10 per MWh congestion fee. How these fees are structured is somewhat artificial. Either way, the users at C end up paying $40 per MWh for electricity, generators at A get $20 per MWh, and generators at B get $30 per MWh. This type of pricing also encourages new generators to locate close to customers to avoid congestion charges.

Setting congestion prices at each node where power is generated or offloaded to large industrial users or local distribution networks is known as *nodal pricing* or *locational marginal pricing.* Calculating nodal prices in principle is quite complicated. There are thousands of locations where power is generated or offloaded, each of which needs a nodal price when some lines are congested. Each of the thousands of links between these points has different physical properties that determine how electricity flows through their lines and thus what the nodal prices should be. Changes in demand, generator shutdowns, and other grid failures (e.g., from lightning) affect, perhaps on a minute-by-minute basis, which if any lines are congested and which are not and thus what the nodal prices should be.

Despite this complexity, many industry participants, advisers, and regulators believe that there are enough data and computing power available to set nodal prices effectively and reliably. The process of calculating nodal prices and the decisions of generation companies to adjust power outputs in response to them can be automated, allowing successful implementation of nodal pricing. For example, FERC approved the use of nodal (or, equivalently, locational marginal) prices in the Pennsylvania–New Jersey–Maryland region, where it has been implemented since April 1998.

Assuming computational and implementation problems are satisfactorily addressed, two problems remain. First, what should one do with the money? Congestion payments may represent a fairly large amount of revenue for the transmission grid. Because the transmission provider is regulated, one could allocate the money raised by congestion fees toward covering the fixed costs of transmission, thus reducing fixed, usage-based, or distance-related transmission fees overall. To the extent that there is money left over, one could return it to users or generators, or cut charges by a constant per-kilowatt-hour amount throughout the grid to get revenues back in line with costs.

Second, how can we encourage transmission operators to expand capacity and ease congestion? Nodal prices suggest where it would be most valuable to expand the grid. High fees can indicate areas that are at either the beginning or the end of congested lines. Letting the transmission company keep the fees, however, would reward them for keeping the network congested.

One potential remedy for this problem, which has been employed in New York, is to auction off the rights to recover congestion revenues to independent entities that do not have operational control over the grid itself. Putting congestion profits in the hands of those who cannot affect the grid eliminates an obvious incentive for the grid manager to keep lines congested. In addition, a market in rights to claim these rents can provide a way for generators to hedge against the risk that congested lines may reduce their ability to sell power. States that adopt this method will need to take care not to create a politically powerful interest group of congestion-right owners that would have an incentive to discourage expansion in order to keep congestion fees high.

It will be the task of the regulators to encourage the transmission operator, either a regulated transmission company or an independent system operator (see Chapter 10), to expand the grid to reduce congestion when the benefits of reduced congestion exceed the costs of construction and maintenance. Providing appropriate institutional or financial incentives to undertake this expansion creates an inherent conflict between keeping rates low (to encourage efficient use of the grid) and keeping them high (to provide rewards for expansion). Achieving the right balance will likely preoccupy state and federal regulators for some time to come.

CHAPTER

Encouraging Competition

The belief that opening retail electricity markets will lead to lower prices and better service for households, offices, and industrial users is predicated on the belief that such markets will be competitive. Such markets may fail to be competitive if only one or a small number of firms supply power to a particular area, or if the power producers agree among themselves not to compete. As we observe an industry in flux, with numerous mergers, divestitures, entrants, and volatile prices, how to ensure competition becomes an ever more pressing question.

The antitrust laws are the main legal means of ensuring that competitive markets remain that way. Because those laws are not designed to control markets, such as electric power, where monopolies arose as a matter of prior regulation, a first policy step in some states could be to require divestiture of power plants to increase the number of independent competitors.

One concern, suggested by the California electricity crisis, is that generators may unilaterally find it profitable to withhold output to raise prices, even when the markets appear competitive by conventional structural indicators. Such concerns have been behind calls for temporary federal caps on wholesale prices. Evidence supporting assertions that market power is being exercised needs to be handled with care. Modifying the operations of electricity markets, or programs to make consumers more sensitive to prices (e.g., installing real-time meters) may reduce the incentive for anticompetitive withholding. If not, then wholesale price caps, particularly during peak periods, could become a permanent feature of "nominally deregulated" wholesale electricity markets.

Mergers among firms that compete could give the firms the ability to raise prices on their own, facilitate collusion among all competitors, or make competition less intense. Deciding whether to block a merger requires understanding which firm competes with which, how competitive the market might be, and which firm might enter the market if the price goes up. In some cases, mergers between a generation company and gas companies could cause problems if the gas company is a primary supplier to the generation company's competitors. Finally, although the industry is in transition, merger evaluation could be so speculative that antitrust authorities may have too hard a time proving that a merger may be harmful.

When the fixed costs of providing a service are very large relative to costs related to how much one produces, a market might inevitably have just a single supplier. As we saw in Chapter 6, the wire sectors of the electricity industry—long-distance transmission and local distribution—are in just that category. When faced with a so-called natural monopoly, particularly over crucial goods and services such as electricity distribution, the policy response may be to regulate prices and terms of service directly rather than to leave such matters to an inadequately competitive market.

However, the potential for exercising market power is not generally preordained by the nature of technology and cost. It can be the artificial result of intentional actions taken to create market power, where competition would otherwise be sustainable. The considerable effort to give customers the ability to choose their power company is predicated on the idea that markets in generation and retailing will sustain enough independent suppliers to be competitive. With competition, electricity producers and markets will have the incentive to come up with cost savings and new marketing ideas. The benefits of those innovations will be passed on to consumers in the form of lower prices and better service.

All of this effort may provide little benefit to consumers if generators retain the ability to keep electricity prices high, particularly if they end up higher than they would have been if markets had been kept regulated. The incentive and ability to withhold output in order to set prices above production costs, referred to as *market power*, is the predictable outcome (roughly speaking) in three situations:

- A single firm faces insufficient competition from other suppliers, leaving it with an effective monopoly.
- Having too few independent suppliers in a market, where the lion's share of the business is concentrated in just a small number of firms, dampens competitive forces.
- Firms in an otherwise competitive market mutually agree not to compete with each other.

Antitrust Laws: Responses to Market Power

In the United States, the primary policy tools for keeping markets competitive are the federal antitrust laws (see box on next page). The primary responsibility for enforcing them rests with the Department of Justice's Antitrust Division and the Federal Trade Commission. Regulatory agencies, including the Federal Energy Regulatory Commission (FERC) and state public utility commissions, have the authority to take competitive effects into account, particularly regarding whether or not to approve mergers among firms in the industries they oversee. State governments can bring suits under both federal antitrust laws and parallel state statutes. Last and hardly least, in terms of the day-to-day operations of most firms

The Antitrust Laws

The antitrust laws are directed at trying to prevent the *acquisition* of monopoly power through collusion, monopolization, or merger. Most U.S. antitrust laws follow from three very brief provisions in the Sherman and Clayton acts, matching the three concerns listed in the text. Section 1 of the Sherman Antitrust Act, enacted in 1890, addresses collusive agreements "in restraint of trade":

> Every contract, combination in the form of trust or otherwise, or conspiracy, in restraint of trade or commerce among the several States, or with foreign nations, is declared to be illegal.

Section 2 of the Sherman act makes it illegal to "monopolize" a market:

> Every person who shall monopolize, or attempt to monopolize, or combine or conspire with any other person or persons, to monopolize any part of the trade or commerce among the several States, or with foreign nations, shall be deemed guilty of a felony[.]

Finally, Section 7 of the Clayton Antitrust Act, passed in 1914 and substantially amended in 1950, addresses mergers, by prohibiting acquisitions by one person or corporation of stock or assets in others,

> ... where in any line of commerce or in any activity affecting commerce in any section of the country, the effect of such acquisition may be substantially to lessen competition, or to tend to create a monopoly.

The vagueness and brevity of the antitrust statutes has kept them applicable despite vast changes in technology, industry structure, and our understanding of the relationships among business practices, market outcomes, and consumer interests. During a century of court decisions, legal scholarship and economic analyses have elaborated, with ever greater detail and complexity, these seemingly simple statements regarding "restraint of trade," "monopolize," and "lessen competition."

and the courts, private parties can also bring civil antitrust suits to block mergers or halt practices they regard as anticompetitive and also to seek damages from firms that may have injured them in violating the antitrust laws.

The reach of the antitrust laws is not unlimited. They are designed primarily to prevent situations from developing that would lead to the exercise of market power. This includes rendering cartels and collusion illegal, preventing strategies designed to impose monopolies on otherwise competitive markets, and blocking mergers that would make the exercise of substantial market power more likely. They are not designed to deal with the exercise of market power by a firm that has a legally acquired monopoly or large market presence. Over the years, the antitrust laws, particularly decisions under Section 2 of the Sherman Antitrust Act, have addressed activities where a monopolist might use its market power to foreclose competition. Such foreclosure could be an act of illegal "monopolization," but because it involves a firm's acquiring a monopoly in new markets rather than in exploiting market power where it already has a monopoly.

To deal with firms that are able to set high prices, typical options are to regulate firms' prices, as with the traditional utilities (telephone service, electricity distribution) or to have the government itself provides the service (water). A more frequent tactic is to wait for firms that are attracted by high monopoly prices to enter the market and eventually to provide the competition that brings prices in line with cost. Fortunately, with some notable exceptions, few markets are inherently immune from competitive forces for a long period. Opening generation markets to competition is predicated on the belief that it is not one of those exceptions.

Monopoly at the Starting Gate

One initial problem in implementing electricity competition is that most states start off with the incumbent utility dominating the market. In and of itself, this need not lead to deregulation without competition. Suppose that consumers in a state where a single utility owns all the generators can obtain power from generators in another state at competitive prices, and that the transmission lines between the states have ample capacity to deliver that electricity. Competitive pressure from those outside generators will keep the in-state monopoly from exercising much market power (see box).

However, leaving matters to the market may not work. Transmission lines may lack sufficient capacity to bring in much power from outside the state. Out-of-state generators may not be able to supply enough power at reasonable prices to counter any attempt by the in-state monopolist to drive up prices by cutting back on power availability. If the market will not keep prices at competitive levels, and the antitrust laws offer no relief, then divestiture of generators by the incumbent utilities may be required so the retail electricity business does not begin as a monopoly. For example, California required incumbent utilities to divest 50% of their generation capacity before the opening of retail power markets; they ended up spinning off all of their generators except hydroelectric and nuclear power plants. Some legislative proposals would clarify that the Federal Energy Regulatory Commission has the ability to devise remedies for undue market power, including possible divestiture of generation units.

In many states, the incumbent utilities have sold their power plants on their own initiative. According to the Energy Information Administration's recent report, *The Changing Structure of the Electric Power Industry 2000: An Update*, investor-owned utilities have, since 1997, divested or are in the process of selling 22% of the total generation capacity in the United States. Such divestitures may allow these utilities to avoid the accusations and scrutiny that might follow if they continued to compete with independent generators while owning the transmission and distribution lines those competitors need (as was described in more detail in Chapter 7). In addition, if designed properly, such divestitures could reduce *concentration* (essentially, how much of the business in a market is concentrated in the hands of just a few sellers) and increase the prospect for competition among generators. When a utility sells all of its generators to the same company, as has happened in some states, the divestiture may do little if anything to alleviate concerns about market power.

"Marketing" Market Power versus Generation Market Power

Selling off generation may not suffice if the utility remains the sole marketing agent for the customers that it served before the opening of retail electricity markets to new competitors. Buyers may be reluctant to switch simply because of relations built up with a familiar utility through decades in which that utility held the monopoly over selling electricity. If an established utility's brand name in electricity is valuable—and other generators cannot break into the market even if they own generators and have equal access to transmission and distribution grids—the incumbent utility may retain the ability to charge a premium for its power at retail. It may have this power even though a divestiture of generation plants may mean that it has no ability to sell electricity to power marketers at wholesale at prices systematically above cost.

What, if any, policy response is called for by this brand name advantage is problematic. Because the benefits of the established brand name are in large part the result of investments made during the regulated era when revenues covered costs, customers have already paid for those advantages. In effect, allowing the incumbent firm to keep those benefits produces a *cross-subsidy* (see Chapter 7) in which past and present ratepayers would be covering the costs of future marketing advantages.

A *divestiture* (i.e., selling off) of the customer base to other marketers, or prohibiting use of the utility's regulation-era brand name, might be warranted. However, the established brand name may provide legitimate benefits to consumers by reducing the cost of finding a power provider with which they feel comfortable. Diluting that brand name, or forcing some consumers to get their power from a company that they have never heard of, could do more harm than good. If so, it may be better just to let generators establish their brand names as retail competition develops.

Competitive Prospects

Even without the legacy of a dominant utility, electric generation markets present some unique possibilities regarding the ability of single firms to charge high prices, particularly during peak demand periods. When demand for electricity is great in a geographic market, it can become extremely costly to provide additional power into that area. If the amount of power supplied were to fall, perhaps even a small amount, prices might have to increase substantially to make up the difference.

Moreover, the demand for electricity is relatively insensitive to price—what economists call *inelastic* (see box). At peak demand periods, when supply is relatively unresponsive to higher prices as well, a firm that has multiple generators serving an area might find it profitable to shut one of them down to drive up the price. It may be profitable simply to cut back operations on a single large generator. Economic models of markets characterized by oligopoly (just a few sellers acting strategically to raise the market price) or a dominant firm (one that sets the price its

Elasticity and Price–Cost Margins: Measuring Price Responsiveness and Market Power

Sensitivity of demand and supply to price are crucial in assessing the likelihood that firms will find it worthwhile to raise prices by withholding output. If buyers are highly responsive to price, sellers will find that they would have to take a great deal of supply off the market to raise the price substantially, reducing the profitability of doing so. Conversely, if buyers keep buying when the price goes up, raising the price is more likely to be profitable.

Similarly, if a single supplier is thinking about raising price, it is interested in how its competitors will react. If its competitors would sell a lot more at a higher price, it is left with fewer sales and lower profits. If the competitors would not increase sales in response to high price—because they are already producing at the limit of their capacity—that single supplier is more likely to find it profitable to raise price.

The standard measure economists use for price sensitivity is *elasticity*. On the buying side, the *elasticity of demand* is the ratio of the fraction by which sales fall to a given percentage change in price. For example, if a 10% price increase would bring about a 20% drop in sales, the elasticity of demand is 20 divided by 10, or 2. If sales would fall by only 2% following this price increase, the elasticity is 2 divided by 10, or 0.2. (Sometimes negative demand elasticity numbers, –2 or –0.2 in these examples, are used to reflect the fact that when the price goes up, purchases fall.) In a similar fashion, the *elasticity of supply* is the ratio of the fractional change in output following a specified percentage change in price.

Using ratios of percentages has many advantages, but the most important may be that we do not have to worry about the way we measure price or output. A 10% increase is a 10% increase, whether we measure output in kilowatt-hours or megawatt-hours, or whether we measure price in cents or dollars.

The effect of elasticities on market power depends on the circumstances under which it might be exercised. A standard measure of market power is the *price–cost margin*, which equals the fraction by which price exceeds *marginal cost* (i.e., the cost of producing the last unit of output sold in the market). For a monopoly or cartel, price–cost margins are inversely related to the elasticity of demand (i.e., the lower the latter, the higher the former).

When firms of roughly equal size choose output independently but in strategic recognition of the expected output decisions of others, the price–cost margin will be roughly the inverse of the demand elasticity times the number of firms in the market. With more firms or a higher elasticity of demand, prices are lower (all else remaining equal). A firm whose competitors are likely to sell more as prices rise will tend to set prices higher the lower is the elasticity of demand, the lower is the elasticity of supply, and the greater is its share of the overall market.

competitors follow) suggest that prices could be double the competitive level in electricity markets at peak periods, even if the market looks competitive by conventional standards (e.g., it has 7–10 or more substantial independent suppliers).

As was stated in Chapter 5, an assertion in discussions of the 2000–2001 California electricity crisis is that it was the result of just this sort of exercise of market power by firms serving California, either through strategic reductions in supply or generator outages. Numerous empirical studies have found very high markups of price over cost in electricity prices, even taking into account increased natural gas costs and environmental regulations.

Such studies, though important and informative, are not yet conclusive. On the price side of the equation, one has to keep in mind that it is not the price one sets, but the price one gets. The prospect that those who bought electricity at wholesale prices would go bankrupt suggests that prices may have been inflated to take the risk of nonpayment into account.

More important differences lie on the cost side. The prevailing method for measuring price–cost margins (see the box on the previous page) or markups is to compare price with a measure of the average variable cost of the highest-cost producer in the market. For markets with excess capacity, this approach is reasonable. However, prices during peak periods have to be sufficiently high so that generators coming on line to meet that demand cover not only their operating costs but their capital costs as well. If so, a high price–cost margin, where "cost" is measured by average variable cost, would be consistent with competition.

An illustration of this principle would be a resort hotel. Its room rates during the peak tourist season are set far above its operating costs to get the revenue necessary to cover the costs associated with the construction and financing of the hotel. In the electricity generation context, a plant that is online only 1% of the time (80–90 hours a year) to meet extreme peak has to recover all of its capital costs 100 times more quickly than a baseload plant running all of the time. Opening markets may only be revealing the high cost of peak power, a cost that remained hidden when it was averaged into regulated rates.

Questions about the empirical basis for claims of market power do not prove that it was not exercised. As was noted above, it would not be surprising if generators unilaterally found it profitable to reduce output, at least in peak periods. But these questions do suggest that policymakers should be cautious in intervening in wholesale power markets to deal with alleged market power.

Setting Wholesale Price Caps

In June 2001, FERC adopted so-called soft caps. When reserve generation capacity is less than 7% of supplies, generators would have to sell power below a calculated "market clearing" price based on FERC's estimate of the "marginal cost of the last unit dispatched," or justify prices charged above this level. When reserve capacity is greater than 7%, prices would be capped at 85% of the price FERC calculated during the most recent preceding supply crunch. In addition, FERC has ordered generators, including publicly owned utilities (see Chapter 13), to offer all the nonhydroelectric power they can produce, either through prior contract or by bidding it into the spot market.

A second potential product market complication involves ancillary services (see Chapter 2). Suppose that a grid operator typically turns to one of three or four power providers to obtain rights to adjust power output—"regulation"—to ensure that loads remain balanced on a minute-by-minute basis. Then, two of those providers announce plans to merge. Whether that merger should be stopped depends on what would happen if, following the merger, these providers attempted to raise the price they charge the grid operator for the right to regulate their power output. If other generators could offer this service as well, "regulation" would be in the same overall market as "electricity," and the merger would be less problematic. If not, the merger might be challenged.

The major geographic question is whether distant power producers would export electricity into an area where local generators attempted to raise prices following a merger. One consideration is transmission cost. The costs of transmission are the rates themselves, and power losses that occur along the way as some of the electrical energy is lost in the form of heat. The larger are the transmission costs, the smaller will be the geographic markets. If markets are local, a merger in electricity (or any other industry) can threaten competition even if the nation as a whole has numerous suppliers.

A potentially significant consideration, especially during times of peak demand, will be transmission line congestion. When generators over a wide area are pumping electricity in large quantities, parts of the transmission grid may no longer be able to carry additional power. Congestion can create *load pockets*, areas in which the only source of additional power would have to come from local generators. A merger of power producers within a load pocket could have serious implications for prices during peak periods, when the ability of users to switch to power from sources outside the load pocket is attenuated by the capacity limits of the transmission system.

Do the Relevant Markets Appear Competitive?

The next step in assessing mergers is to look at concentration. Intuitively, the more concentrated a market, the less competitive it will be, and the more the largest firms will be able to raise prices on their own or fix prices among themselves. The building block in looking at concentration is *market share*. Frequently, it is easier to measure a firm's share by looking at the percentage of sales or revenue it has of the total market. If the data are available, capacity is often a better measure, because it is more closely related to a firm's ability to increase sales in response to an increase in price. In electricity, generation capacity is typically relatively well known, and should be the basis for deciding when a merger leads to undue concentration.

Before the early 1980s, the main method for measuring concentration was the "*N* firm concentration ratio," the aggregate share of sales or capacity of the *N* largest firms in a market. Since 1982, the preferred method has been the *Herfindahl-Hirschman Index* (HHI), which is the sum of the squared market shares of all the firms, with 10,000 (100 percent, squared) describing a monopoly market and 0 the limit of a market with no individual seller having an appreciable share. For example: If four generation companies currently supply a market, and their shares are 40, 30, 20, and 10, then the HHI will equal $40^2 + 30^2 + 20^2 + 10^2 = 1{,}600 + 900 + 400 + 100 = 3{,}000$.

The HHI tends to give more weight to big firms (e.g., an industry with one big seller and two tiny ones will have a bigger HHI than one with the same number of sellers of roughly equal size). As a rule of thumb, the *Guidelines* tend to regard industries with an HHI below 1,000—what one would get with 10 competitors of equal size—as sufficiently competitive to make mergers not worth worrying about. As the HHI increases, particularly beyond 1,800, mergers—especially those between the larger firms in an industry—are more likely to catch the attention of antitrust enforcers.

Measures of concentration are very sensitive to the defined relevant market. Consumers in a particular area may be able to choose power from a number of producers most of the time, and measured concentration may be quite low. But those same consumers may be very limited in their choices during peak periods, when transmission lines may be congested and only a few generators are capable of meeting demands. Markets at peak periods in load pockets may be very concentrated, and a merger that is otherwise benign could substantially raise power prices at particular times and places.

Competitive Effects

Merger analysis is designed to deal with the three circumstances leading to market that we identified above. Under the *Guidelines*, it is not enough to identify the relevant market and show that it is concentrated. Those contesting a merger also should identify the competitive effects (e.g., what exactly they think is going to happen if the merger goes through).

The *Guidelines* partition the set of competitive effects into two categories. The first, labeled *unilateral effects*, deals with the first two settings outlined at the beginning of this chapter, both of which involve decisions firms make on their own. The merged firm itself would be large enough to find it profitable to raise prices without having to worry about losing business to its competitors. In addition, by reducing the number of independent competitors, a unilateral-effects merger may reduce the intensity of competition, leading to higher prices throughout the market, without necessarily giving any one firm inordinate market power or enabling explicit collusion. The likelihood of either of these possibilities tends to rise, all else equal, the larger is the market share of the firm after the merger, and the smaller are the elasticities of demand from buyers and supply from competitors.

The second category, explicit collusion among erstwhile competitors, is termed *coordinated effects*. This conforms to the last of the three potential market power settings outlined at the beginning of this section. A case based on coordinated effects would require a description of how a merger would make collusion more likely to succeed. This might include how firms that attempted to cheat on the agreement by selling at a discount would be detected and punished, without the ability of law enforcement agencies to detect and prosecute the cartel under the antitrust laws.

Entry: The Crucial Factor

Neither unilateral exercises of market power nor even explicit collusion need lead to higher prices if firms not in the market would enter in response to higher prices. But entry into a business takes time. Moreover, entry requires making initial expenses in production and marketing. If the volume of business an entrant would be likely to get is not large enough to generate enough revenue to cover these and other expenses—especially recognizing that the entry itself will tend to reduce prices—then entry would not be likely. Consequently, a merger would be more likely to lead to higher prices. For purposes of analyzing mergers, the *Guidelines* recommend disregarding entry as a competitive force if it would take longer than two years.

If new generation can be built relatively quickly in response to a monopoly, or if transmission capacity can be expanded to increase the imports of power into a specific geographic area or load pocket, a merger between two major producers should cause less concern. However, expansion of generation or transmission may not be easy, particularly if local land-use policies inhibit the construction of these large industrial facilities. Also, new generators are unlikely to enter unless they think they can make money, even if the price goes down after they add their competitive presence to the market. They may not build facilities to serve a high-price state unless they believe they can profitably sell electricity in other states as well. Accordingly, transmission capacity out of an area may be an important factor in regarding new entry as a threat to market power.

Finally, we note the potential for on-site production of electricity, referred to as *distributed generation*. If generators can be installed in factories, office buildings, and homes, then mergers among traditional utilities or independent power producers may not cause concerns. Were they to attempt to raise prices, electricity customers might find it worthwhile to produce power for themselves. Distributed generation may not be quite that economical yet. But as we noted in Chapter 6, improvements in that technology may not only put competitive pressure on large-scale generators, but it could also eliminate the need to regulate the transmission and distribution systems that are required to deliver energy to the users' premises.

Might a Merger Make Matters Better?

A merger might improve performance, first by reducing costs and increasing efficiency, and second by redeploying assets that might otherwise disappear because a firm is leaving the market. In theory, these benefits could counteract the anticompetitive effects of a merger, but in practice rarely seem to do so.

■ *Efficiencies.* A merger in a concentrated market that would lead to higher prices, and where entry would not serve as an adequate remedy, could still generate net benefits to the economy if it led to cost savings. In theory, a small percentage reduction in costs can outweigh the economic harm of higher prices. The benefits of those savings would be felt over all production, whereas the costs of market power are based only on the volume of sales that is lost because prices have gone up.

Nevertheless, antitrust authorities tend to be skeptical of efficiency justifications for otherwise harmful mergers. Efficiency claims are easy to make but hard to support. For example, it may well be that one big firm can produce at lower costs than two smaller ones, but a merger does not magically transform the established plants and operations of the two small firms into those that a larger firm would have built from scratch. Also, a merger may not be necessary to realize these benefits; the two small firms could compete independently to get bigger rather than eliminate competition among themselves. Finally, to some extent authorities may count benefits to consumers more than savings to producers. If so, they would give efficiencies less than full weight, even in the rare cases where they are both incontrovertible and achievable only through the merger, unless the parties can show that they would be "passed through" to consumers in lower prices.

■ *The "failing-firm" defense.* Mergers may be justified when one of the firms is failing. If a firm would be going out of business in any event, it would seem to offer no competitive potential that would be preserved by blocking the merger. Before concluding that the failing-firm defense should apply, authorities will typically ask first whether there is a less anticompetitive alternative (e.g., purchase by a less competitively significant participant in the market or a buyer not currently in the market at all).

Also, it is not enough that the firm is failing; its productive assets would need to effectively disappear. Suppose a firm were to go out of business, but its factories would not be demolished or irreversibly modified to produce different products. Keeping those factories out of the hands of other firms in the market could preserve some competitive pressure, because if prices rose, someone else could use those factories to enter the market.

Vertical "Convergence" Mergers

Most of the concern regarding mergers in the electricity industry should be focused on horizontal acquisitions that might reduce direct competition between generators in the sale of electric power. Vertical mergers, in general, involve operations that go between markets, and thus do not change concentration and other determinants of the competitive environment within markets, whether it already favors competition or monopoly. Two firms that frequently deal with each other on a buyer–seller basis, or that offer products that consumers often purchase together, may find that they can better coordinate their design, fabrication, and marketing processes as one firm than they can as two. In addition, if firms with market power at vertically related stages in the production chain merge, the predictable result is lower prices and more sales. After a merger, each firm would take into account how raising prices would hurt the vertically related partner, and thus will have less incentive to do so.

However, vertical mergers—between generation companies and firms operating at different levels in the industry—can in some cases raise concerns that regulators and antitrust authorities may address. One type of merger that could cause a problem would be one between a generator and a regulated distribution or transmission company that delivers electricity from it and its competitors. Such a merger could give the regulated "wire" company the incentive to discriminate against these competitors and favor its affiliated power company in terms of access, service quality, or (if the distribution companies also manage markets; see Chapter 10) the ability to sell their power. Such discrimination could make these competitors less effective, causing power prices to increase (Chapter 7).

A second kind of vertical merger is that between an electric generator and a supplier of fuel used to generate electricity, particularly natural gas. Such mergers are

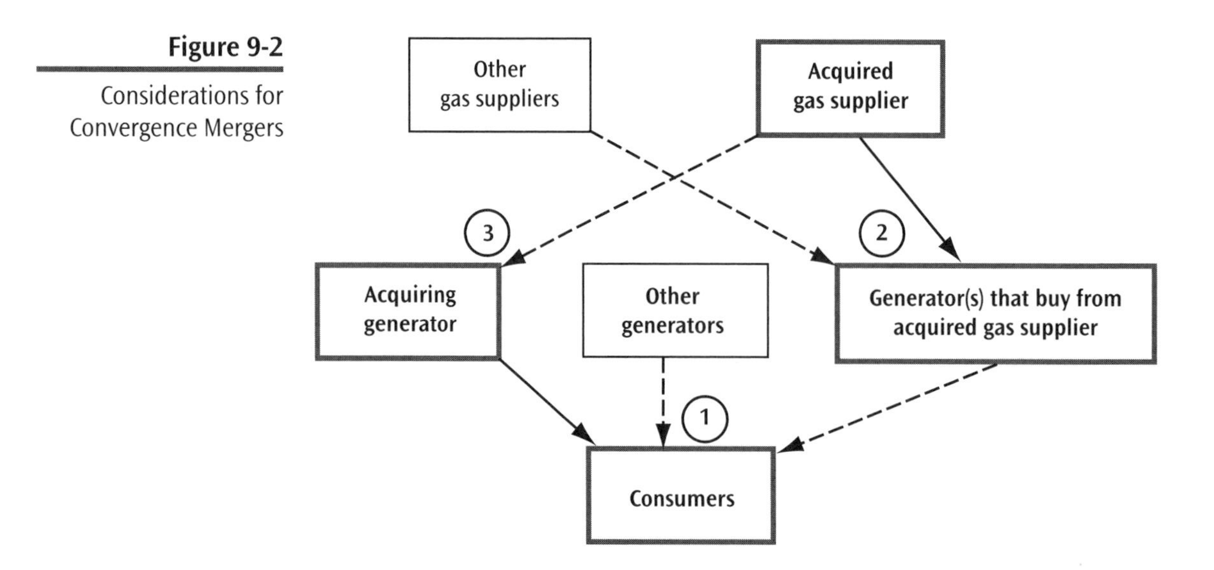

Figure 9-2

Considerations for Convergence Mergers

called *convergence mergers*. Figure 9-2 illustrates a convergence merger and how it might create competitive concerns.

Suppose that we have a merger between the acquiring generator and the acquired gas supplier, as is shown in Figure 9-2. A first question in looking at this convergence merger would be whether a direct horizontal merger between the acquiring generator and the generators that buy from the acquired gas supplier would hurt consumers (circle 1). If the answer is yes, the next question is whether the acquired gas supplier effectively exercises control over the output of its customers, most obviously by exercising market power over them (circle 2).

If the answer to this question also is yes, we then should ask if that gas supplier already controls the output of the acquiring generator (circle 3). If so, the convergence merger is perhaps paradoxically less troublesome, because the acquired gas supplier already effectively controls output and price in the downstream electricity market. A convergence merger is more likely to be troublesome when the acquiring generator and acquired gas supplier have no prior dealings, for example, if the acquiring generator is fueled by coal or nuclear power.

Combining before the Dust Settles

We end the chapter with a specific problem with merger assessment during this transition period from closed to open markets. The extent to which markets are competitive (i.e., the extent to which generators in one location will face competition from generators 10, 100, or 1,000 miles away), depends on how a wide array of public policy questions are settled at both the state and federal levels. These include, among other things, the magnitude and structure of transmission rates (see Chapter 8), how the industry and markets will be organized and run (Chapters 7 and 10), and the lifting of present federal laws that impede retail sales by multistate utilities (Chapter 12).

Yet, before these questions have been settled, utility mergers continue to take place and must be assessed. Under antitrust law, the burden of showing that a merger should be stopped rests with the government. But during the transition period, before retail competition takes place and the rules have been worked out, the government may not have the evidence to prove that the two firms will be competitors in the near future. Furthermore, there may still be some unsettled rules of the game (e.g., transmission pricing policies) that could inform judgments as to which firms are likely to compete with which. The only market experience may be from the precompetition era, when utilities operated as monopolies in separate areas, were not in the same market, and hence did not compete.

Accordingly, the government may have no market evidence to show that a merger will lead to higher prices. Ironically, we might expect that the decision of two utilities to merge could well be affected by their desire to do so before the authorities are able to keep them from exploiting market power after retail competition begins. Thus, mergers could be more likely to be anticompetitive just when authorities are in a poor position to contest them under general legal standards. Whether it would be better to devise transitional policies on the timing and possibility of utility mergers before the resolution of policy questions that affect market definition, or before a record of actual market experience can be established, could prove to be a vexing question. The volume of mergers undertaken and approved so far, however, suggests that the horses may have already left that particular barn.

CHAPTER

10

Balancing Loads and Dispatching Power

Electricity stands out in that, unlike virtually every other commodity, disaster can strike unless producers supply exactly the amount that people want to buy at any given time. Keeping power production and use in line—load balancing—will require the active involvement of generators, transmission companies, local distributors, and customers, as well as the regulators that oversee the industry. In implementing electricity restructuring, policymakers should consider how to guarantee the provision of ancillary services needed to keep loads balanced on a minute-by-minute basis and to provide emergency power when generators or transmission lines unexpectedly fail or demand is unexpectedly great.

A first question is whether each generation company should be responsible for keeping its own power supply in balance with its own customers' desires. Because failure to meet power demand causes a breakdown of the system as a whole, and not just a blackout to that company's customers, letting the market take care of it may not suffice. Generators may need to meet standards for maintaining power and having reserves available, or they may need to be held liable when their inability to meet demand brings down the larger grid. If those prove inadequate, distribution and transmission companies may need to take on the responsibility of providing ancillary services and holding power in reserve.

Involving grid operators in the business of maintaining loads has led many states to involve them in the overall management of power markets, through taking bids from producers and users and dispatching generators as needed. The grid need not be involved in this aspect to control generation costs; the electricity market, like any other, can handle that through letting generators compete for customers. But whether such a market is compatible with keeping loads balanced and systems secure is perhaps the crucial question facing electricity policymakers.

Electricity is an unusual commodity in many respects. One is that the power produced by a particular generator does not go to that generator's customers. More precisely, if a generator sells *N* kilowatts of power to its customers, it is merely committing to inject *N* kilowatts of electricity into the overall electricity system at the same time that the customers are taking *N* kilowatts out of it. It is as if Starbucks sold *M* cups of coffee by dumping that volume of coffee into a common vat mixed with coffee from every other coffee shop, out of which its customers had the right to pour out *M* cups.

As a consequence, the distinction between the central system (i.e., the "grid") and the power pooled within it can become blurry. If coffee were sold as in the Starbucks scenario, one might well expect that the owner of the vat—the grid—might find itself becoming involved in the wholesale purchase and retail sale of the coffee within it. This blurriness could be especially pronounced if the vat owner became responsible for the quality of the coffee supplied, that is, for making sure that the caffeine jolt is "reliable." We discuss in this chapter and the next how these considerations have affected and may continue to affect the development and feasibility of competition in electricity markets.

Load Balancing

Perhaps the most crucial feature that distinguishes electricity from other commodities is the need to keep supply equal to demand on a virtually minute-by-minute basis. For most other commodities, buyers can wait a bit if the item is not on the shelf, or the line on the telephone is busy. Sellers sometimes may have to backorder items not on the shelf, or keep inventory around a little longer when items do not sell as fast as expected. Both of these can be costly and inconvenient, to be sure. But they are not catastrophic in the way that a mismatch between electricity demand and supply can be. If more electricity is demanded than generated, brownouts or blackouts follow. If more electricity is supplied than used, the heat from the extra energy can damage the transmission and distribution systems.

Keeping electricity supply just equal to demand, by varying either production or use, is called *load balancing*. Two properties of electricity exacerbate the problem of keeping loads balanced. First, the cost of storing electricity in substantial quantities is prohibitive. If one is worried that the store will run out of soup or toilet paper, one can keep some spare supplies in a cupboard or closet. A seller that thinks that demand could be stronger than expected can keep an inventory of the commodity available on the shelves or in the warehouse. Neither tactic is available for electricity. Batteries are too expensive to store much power for most users, and at least up to now, generation on-site is prohibitively costly for all but large factories or commercial facilities that cogenerate electricity as a by-product of energy available from other production processes or space heating systems. Hence, when users want electricity, the generators have to be producing it at that moment, and the transmission and distribution grids have to be able to deliver it.

The second problem is that load imbalances take down the entire system or entire regions within the system, not just those who are customers of a particular power company that happens not to be producing enough to meet their demands. The power produced by everyone essentially becomes part of a common pool from which all users draw. If what is there does not suffice, all customers on that grid lose, even if the cause of the insufficient supply is a failure to produce by one generator or unanticipated demand from just its customers.

Unless or until technology becomes available with which a distribution system can rapidly cut power to users of a particular generation company, failure to balance loads by one company potentially harms all. Similar problems occur with producing more power than buyers use at any given time. Moreover, unless the power usage of each customer can be monitored on a minute-by-minute basis—as opposed to reading the meter once a month—it may be impossible to know whether a particular generator is supplying less or more power than its users need.

In short, the inability to store large amounts of electricity means that supply must constantly be kept equal to demand. The systemwide nature of the effect means that the costs of failing to keep loads in balance are borne by everyone on a grid and not just the company that happens to be out of balance. Accordingly, a laissez-faire attitude may not work. At a minimum, generators that get out of balance with respect to their customers' loads have to be responsible for procuring the additional power or cutting back production, as necessary, to restore the appropriate balance. Providing an incentive to ensure such responsibility means, in some way, holding generators liable for the costs imposed when they get out of balance.

If these incentives do not work, the responsibility to ensure that loads are balanced falls to the relevant grid manager—the local distribution company and, perhaps, the regional transmission organization, either an independent system operator or a transmission company managing the transmission system (see Chapter 7). As we will see below, this creates an ongoing role for these regulated wire monopolies in ostensibly competitive power markets. Moreover, the need for load balancing may lead wire operators to go beyond procuring power and network management to ensure balanced loads. In some states, wire operators have begun to manage the overall market for power and, in effect, continue to exercise the authority over the dispatching of power that they held before the advent of restructuring.

Ancillary Services: What It Takes to Balance Loads

A full understanding, and perhaps even an adequate idea, of what it takes to keep the grid balanced and functioning requires a detailed knowledge of electric power engineering. But to understand how much (if any) authority to give wire companies to manage the electricity market, one needs a handle on these technical issues.

The term *ancillary services* refers to the category of activities associated with keeping the network going when loads happen to go out of balance (see the box titled "Ancillary Services" in Chapter 2). Typically, these services are procured in advance by the grid operator, which involves payments for being available to provide these services on call and for any energy that is actually produced or taken off the market to keep production equal to consumption. How one lists and classifies the services that fall in this category seems to be a matter of some difference, if not controversy, among those familiar with grid management. However, at some risk of oversimplification, we can point to three major types of ancillary services: regulation, load following, and emergency power replacement.

Regulation

In principle, every time one turns on a light, appliance, computer, or other device powered by the electricity grid, more power has to be added to meet the load. When the devices are shut down, the supply of power has to be reduced. In practice, the effects of any single device being turned on or off on the overall voltage of the system, holding constant the amount of power generated, are within the tolerance levels of the grid as a whole and the devices attached to it. Moreover, things tend to average out (e.g., one household's refrigerator may switch off just as another's comes on).

However, moment-by-moment fluctuations can require that more power needs to be put into the grid, or that generation needs to be cut back. Some industrial users themselves impose substantial loads, and variations in their power use can cause voltage problems in their portion of the grid. Also, sometimes power use decisions made by individuals, though each small, may be correlated and thus create substantial variations that go beyond systemwide tolerances. For example, the onset of clouds may cause homes and offices to turn on lights at more or less the same time. Finally, as more and more devices are either computers themselves or incorporate digital semiconductor controls, overall sensitivity to variations in voltage becomes greater, making it more important to limit any gaps between actual power supplied and nominal power demanded.

As was described above, if the patterns of use over time—*load profiles*—of individual users cannot be monitored on a minute-by-minute basis, a system can suffer brownouts, blackouts, or overheated lines. When these substantial fluctuations occur, the power supply needs to be adjusted to meet the differing loads. The practice of adjusting generation up or down on a per-minute basis to cope with real-time variation in use is called *regulation*—admittedly confusing when it comes up in the context of moving to "competition" and away from "regulation" in the economic sense of the term. Generators supply regulation essentially by offering to make their power adjustable upward and downward on a minute-by-minute basis. They can do this by keeping extra capacity in reserve and by being willing to reduce supply while generating sufficient power to meet minimum operation requirements.

Load Following

If power regulation deals with momentary variations, *load following* deals with variations that take place systematically over a given but short period of time (e.g., an hour). To some extent, load following is an artifact of the way electricity is bought and sold in a particular retail market. Generators may make commitments to supply a particular amount of power during, say, a particular half-hour or hour time period. (Why power may be marketed this way is discussed below.) But during that time period, loads may trend upward or downward. For example, on summer afternoons, electricity demand typically will be greater at 1:59 p.m. than at 1:01 p.m., and less at 6:59 p.m. than at 6:01 p.m.

To keep loads balanced, generators must be around to supply extra power at the end of time periods when demand is increasing, and at the beginning of time periods during which demand will be falling. This service, called load following, need not be supplied by only those generators willing to cede some control over output to meet minute-by-minute load fluctuations. Because load trends are somewhat predictable, generators can compete in advance to provide load following power within the time segments during which most power is bought and sold.

Emergency Power Replacement

Generators—like any other large piece of complicated machinery—sometimes fail. But because the grid is interconnected, the failure of a generator means that the entire system is at risk unless replacement power can be brought up to speed quickly. When the failure is anticipated (e.g., taking down a turbine for scheduled maintenance), the generation company can contract with other power providers to supply its needs during the shutdown period.

Unanticipated failures are more difficult to handle. A partial solution is for grid managers to require all power companies to maintain reserve requirements that can be brought up to speed quickly. Generators can also take payments to hold additional capacity offline that can be made available in the event of a power outage. In effect, the generators charge for the option to call upon their capacity following an unforeseen power failure; they also get the market price of electricity if and when that capacity is used.

Further distinctions for emergency power replacement may be based on how quickly the generator can be *ramped* or powered up. Some power can be brought to bear in 10 minutes; other power may be supplied on the condition that it be available in 30 minutes or more. *Spinning reserves*, capacity from generators already in operation, may be made available very quickly, but it is costly to keep in reserve because of the energy cost in keeping the generator powered up in a stand-by mode. The value of nonspinning reserve capacity depends on how quickly the generator can be brought online. Different technologies can be brought online at different ramp rates.

Who Should Be Responsible for Ancillary Services?

These basic ancillary services of regulation, load following, and emergency power replacement, and others as well, are crucial to keep an electricity system in operation. At least in theory, these services can be supplied through competitive bidding among generators (e.g., in the price they need to be paid not to produce, to keep power in reserve if needed later). The possibility of competitive sales of ancillary services, however, does not tell us whether decisions to buy them can be left to the market. Because power delivery from one generation company to its retail customers cannot be isolated on a power grid, a load imbalance between them affects the stability and reliability of the entire network. Ancillary services are akin to what economists call *public goods*. Because not just a single power company but the entire grid benefits when load balance is maintained, and because not just that power company but the entire grid suffers when it is not, the incentives for that company to secure ancillary services will be too small.

Consequently, maintaining load balance will require some additional intervention. At least four options are available:

- Hold generating companies liable ex post for outages.
- Impose ex ante requirements to provide ancillary services.

- Have the grid operator purchase ancillary services from independent generators.
- Let the grid operator own sufficient generation to keep loads balanced.

Each option has its advantages and disadvantages. We discuss them in turn.

Hold Generators Liable Ex Post

A common way of having persons or companies bear the costs of their actions is to have them cover the costs of any bad consequences. Following this "tort liability" model, one could estimate the cost of not keeping loads equal to power supplies. A generator that lets loads get out of balance with supplies can be held responsible for the costs such an imbalance creates, including the cost of a blackout.

Under a liability model, the incentive to avoid having to pay the cost of blackouts would lead generators to accept the responsibility to provide the ancillary services necessary to keep load balanced. They could contract with other generators for regulation, load following, and spinning and nonspinning reserves. Competitive markets in ancillary services could follow. If the costs of outages is set correctly and ancillary services markets are competitive, the price of those services will reflect both the marginal cost of providing them and the marginal benefit they produce in reducing the likelihood of a blackout.

Leaving ancillary service provision to a market driven by liability rules has a number of disadvantages—over and above the sheer difficulty in deciding accurately how much a generator should pay if its failure to balance loads leads to a two-hour blackout. Liability rules provide good incentives only when a court or regulator can reasonably attribute harm to a particular actor. When the grid is threatened because a power plant has gone out of service, attributing the harm to the company that owns the plant is reasonably clear. With regulation and load following, however, determining that it is generation company X that is out of balance with its customers, rather than generation company Y, may be difficult. Such determinations would require, among other things, that the use of each power company's customers can be monitored at close to real time. In at least one case, opponents of such a proposal have claimed that such a requirement discriminates against new entrants that may be less able to monitor their customers' minute-by-minute use.

A second disadvantage is that liability threats matter only to the extent that a power company thinks it would have to pay if its actions are responsible for a power outage. For large judgments, a power company may find it less costly to file for bankruptcy rather than pay. The bankruptcy option limits the ability of the threat of an after-the-fact liability payment to get power companies to take appropriate action to ensure system reliability. Moreover, bankruptcy procedures, which have evolved to handle corporate reorganizations in markets that can tolerate supply and demand imbalances, are not necessarily designed to promote efficient operations in electricity, where load balancing is critical.

Ex Ante Reserve Requirements

One method to ensure that the power supply and demand are kept in balance is for the grid manager or public service commission to refuse to connect generation companies to the grid, or allow power retailers to operate, unless they have met requirements to procure ancillary services and maintain appropriate reserves.

CHAPTER

11

Ensuring Reliability in a Competitive Market

The U.S. electric power system has had a strong record of uninterrupted service made possible through the cooperative efforts of the utilities that are linked together on the three major U.S. transmission grids. As the electric power industry becomes more competitive, this voluntary approach to ensuring reliability is threatened, while at the same time the transmission system is facing greater stress from more intensive use.

Reliability can be classified in terms of adequacy and security. In a competitive world, the market is expected largely to handle adequacy of generation. However, adequacy of transmission and distribution will still be subject to regulatory oversight. The security of the power system will remain a responsibility of centralized system operators because of the large spillovers associated with the failure of generators or transmission lines.

Restructuring poses challenges for the reliability of both the distribution system and the bulk power transmission system. The threats to integrity and the consequences of failure are greater for the transmission system than for a local distribution grid. To maintain the security of the bulk power transmission grid, power control area operators and security coordinators may need to interfere with the commercial transactions on the electricity grid. However, it may be difficult to distinguish between an action taken to protect system security and one taken for other, perhaps anticompetitive reasons.

Given the potential threats to reliability posed by electricity restructuring, legislators and energy regulators should develop a strategy to protect system reliability as they design and implement policies that will set the course for electricity markets in the future. Such a reliability strategy is likely to include an expanded North American Electric Reliability Council with greater authority, a greater role for the Federal Energy Regulatory Commission in overseeing reliability, and more use of incentives to promote the efficient use of the transmission and distribution systems.

The U.S. electric power system is widely recognized as being among the most reliable in the world. With very few exceptions, the nation's bulk power transmission system—the system that carries power long distances at high voltages—has reliably delivered a continuous supply of electric power that has met consumer and business needs during the past several decades. This strong record of uninterrupted service has been made possible through the cooperative efforts of the utilities that are linked together on the three major U.S. transmission grids. Since the establishment of the North American Electric Reliability Council (NERC) in 1968 following the 1965 blackout in the northeastern states, electric utilities throughout Canada and the United States have worked together to establish and implement voluntary operating rules and procedures to keep power flowing throughout the entire region.

As the electric power industry becomes more competitive, this voluntary approach to ensuring reliability is threatened. As a result of competitive pressures, utilities are facing stronger incentives to cut costs and reduce maintenance expenditures, both of which could also threaten reliability. The tradition of cooperation among utilities in reliability matters is being eroded by the onset of competition. Without the guarantee of cost recovery that existed under rate-of-return regulation, utilities are likely to be more reluctant to make expenditures that could benefit their potential competitors as much as themselves.

These changing incentives come at the same time that the electricity system faces greater stress. The breakup of vertically integrated electric monopolies and open access to transmission have produced a surge in the number of participants in electricity markets. The larger number of market participants has, in turn, spurred an increase in the amount of power being traded and shipped long distances and in the use of the transmission grid to facilitate those transactions. Increased use of the grid means a greater probability of system disruptions. In light of these changing incentives and market realities, promoting reliability will require new policies and institutions.

In this chapter, we define reliability and discuss the activities that system operators, transmission grid owners, and others currently undertake to maintain reliability. Then we detail the threats to reliability that arise with industry restructuring and identify actions taken in the name of maintaining reliability that could stifle competition in energy markets. Finally, we discuss ways to promote reliability in competitive electric power markets.

What Is Reliability?

In engineering terms, *reliability* refers to the ability of the electric system to deliver electric power to consumers with reasonably few interruptions. The reliability of a system is measured by the number or frequency of interruptions during a particular period, the amount of time they last, and the number of customers they affect. The relative importance of these different aspects of power outages depends on the end use of the electricity. Computerized activities—including automated manufacturing and hosting of commercial Internet websites—tend to be fairly intolerant of frequent power outages, no matter how short. Refrigerators may tolerate more frequent interruptions, provided they are short. Outages that affect vast numbers of customers could result in significant disruption and cost, particularly if they last more than a couple of hours.

Reliability can be further classified in terms of adequacy and security. Power system operators and electricity regulators usually use *adequacy* to refer to the capability of the power system to satisfy demand. Historically, the desired level of capacity deemed necessary to meet demand and satisfy reserve margins was centrally determined under the criterion that supply was adequate if the system could be expected to suffer a blackout only once in 10 years.

As electric power markets become more competitive and electricity consumers become more active in real-time electricity markets, adequacy of the generation supply will become an issue for the marketplace to determine. During periods when electricity demand is approaching the capacity limits of the system, electricity prices will rise, and at least some consumers will be able and willing to curtail their electricity use. High peak-period prices will also draw potential new suppliers into the market and give stronger incentives to restore power in case of an outage, which should help to moderate price spikes in the long run. In the short run, however, markets may not be able to ensure adequacy, particularly if retail prices are not allowed to rise to reflect the true cost of power during periods of extreme shortage (see Chapter 5).

Transmission adequacy, conversely, will continue to be subject to regulatory oversight. As was discussed in Chapter 8, transmission remains a natural monopoly with presumably regulated pricing. Although expanded wholesale markets for generation could signal where new transmission lines would be most valuable, state and federal utility regulators will continue to oversee investment by transmission-owning utilities and will be ultimately responsible for making sure that there is sufficient transmission capacity to keep the system operating reliably.

Security refers to the ability of the power system to withstand sudden disturbances, such as the unanticipated loss of a system component. When a major transmission line or large generator is suddenly lost, a regional electric power grid, functioning essentially as one large machine, will experience that loss as a reduction in the frequency at which electricity is supplied to all customers below its standard value of 60 cycles per second or 60 hertz until that component is either restored or the services it provides are replaced by another component. Electric power systems are designed with redundancies and backups to help the system recover quickly from the loss of any component, up to and including the largest generator or the largest transmission line. As the electric power industry becomes more and more decentralized, the security of the power system will remain a responsibility of centralized system operators, owing to the large spillovers associated with the failure of generators or transmission lines.

Systems to Ensure Reliability

Most power outages in the United States are the result of weather-related equipment failures on a local distribution grid. These outages tend to affect a localized group of customers, although they can be spread over a large geographic area and can last a long time, particularly if many local distribution lines have been taken down. However, distribution outages generally affect fewer customers than

Figure 11-1

North American Electric Reliability Council Regions

Note: ECAR, East Central Area Reliability Coordination Agreement; ERCOT, Electric Reliability Council of Texas; FRCC, Florida Reliability Coordinating Council; MAAC, Mid-Atlantic Area Council; MAIN, Mid-American Interconnected Network; MAPP, Mid-Continent Area Power Pool; NPCC, Northeast Power Coordinating Council; SERC, Southeastern Electric Reliability Council; SPP, Southwest Power Pool; and WSCC, Western Systems Coordinating Council.

Source: North American Electric Reliability Council.

the much less common failure of the bulk power system—such as the Western power outage of the summer of 1996, when more than 2 million customers lost power.

Reliability of the distribution system is the responsibility of the local distribution company. Because most of the power outages on the distribution grid are the result of power lines being taken down as a result of extreme weather (falling trees, strong winds, or lightning strikes), one way to improve reliability is to bury distribution lines. Unfortunately, this option is extremely expensive for existing systems, although buried lines are generally the method of choice for delivering electricity to new real estate developments. Smaller expenditures, such as trimming trees around distribution lines and regular maintenance, can also help to improve or maintain the reliability of the grid.

Since it was established in 1968, the North American Electric Reliability Council has been responsible for establishing standards, guidelines, and criteria to enhance the security of the high-voltage electric power grid and to evaluate the adequacy of the power supply, including reserves to handle contingencies. NERC is a voluntary, nonprofit corporation jointly owned by the 10 regional reliability councils identified in Figure 11-1, which in turn are jointly owned and operated by the member electric utilities. NERC has standards governing planning and investment and the operation of the bulk transmission system. The regional reliability councils implement in detail NERC's general standards, to fit the needs of their specific regions. The planning and investment standards are designed to ensure that the system has or will have adequate overall transmission and generation capacity to meet expected growth in demand and accommodate expected power transfers. The operating rules developed by NERC and its regions place limits on the use of high-voltage transmission lines and other key system components to protect the system from contingencies. For example, if a transmission line would be overloaded in the event of a failure of another piece of equipment on the system, then system operators reduce the use of that line to a level that keeps it available to be used in the case of an emergency.

System overseers at two levels help to implement NERC reliability standards. To help ensure that reliability concerns are addressed adequately over large geographic regions, NERC has established 23 security coordinators who monitor large areas and organize emergency responses. Beneath these security coordinators are approximately 150 power control areas (PCAs) within the three major U.S. power grids: the Eastern Interconnect, the Western Interconnect, and the Electric Reliability Council of Texas (ERCOT). PCA operators are responsible for monitoring what is happening on the system in their area, for coordinating power interchanges with other control areas, and for maintaining system frequency at 60 hertz. They take actions and direct others to take action to keep power flowing and to prevent or correct for system emergencies. PCA operation and security coordinating functions are managed by different types of entities—including investor-owned utilities (IOUs), public power utilities, independent system operators (ISOs), regional transmission organizations (RTOs), and regional reliability councils—in different regions of the country. Compliance with operating instructions and NERC rules and guidelines is essentially voluntary; there are no legal sanctions for failures to adhere to them.

Could Restructuring Decrease Reliability?

The restructuring of the electricity sector poses a number of challenges to the reliability of the electric power system. In this discussion, we will consider the integrity of local distribution systems separately from that of the bulk power transmission system. Although restructuring poses new challenges for both parts of the electricity power system, the threats to integrity and the consequences of failure are greater for the bulk power transmission system than for a local distribution grid.

Distribution System Reliability

The reliability of the distribution system is what economists call a *private good,* in that the benefits of expenditures to improve it accrue to the customers of the distribution grid and, presumably, to the owners of the grid. Because these benefits fall to the distribution grid and its customers, regional or national policy to promote distribution reliability generally will not be necessary.

The effect of restructuring on distribution system reliability may depend on whether the local distribution company is vertically integrated into generation or retail energy sales and on how local distribution companies are regulated after restructuring. The effect of restructuring on willingness to maintain the distribution grid may also differ from its effect on incentives to restore power as quickly as possible in the event of a power outage.

Restructuring is more likely to diminish incentives to undertake system maintenance intended to prevent future equipment failure than incentives to make repairs to restore power in the event of an outage. If, as a result of restructuring, distribution companies specialize exclusively in distribution and continue to be subject to rate-of-return regulation, then they probably will not have an incentive to reduce expenditures on system maintenance. As long as they are able to recover the costs associated with maintaining lines and other activities to promote reliability, they will retain roughly the same incentives to do so that they had prior to restructuring.

However, if the distribution company continues to generate electricity or sell electricity to retail customers, it could have a reduced incentive to undertake system maintenance after restructuring. Suppose, for example, that as reported distribution costs go higher, a regulator becomes more suspicious that a utility is engaged in cross-subsidization—that is, in assigning costs of competitive enterprises to its regulated distribution business. If this is so, a distribution utility might have an incentive to cut maintenance and other reliability-related costs, in order to reduce such suspicions and facilitate cross-subsidization. Shifting costs in this manner gives these diversified firms an artificial cost advantage when competing for retail customers. (For more discussion of cross-subsidies, see Chapter 7.)

Restructuring is less likely to reduce incentives to make repairs to restore power in the event of a distribution outage. The distribution utility would lose revenues from payments to use its grid. Sustained power outages would have a very negative effect on public relations for the distribution utility. Utilities also face the possibility of lawsuits from business and commercial customers seeking to recover losses associated with the power outage, and these claims generally would tend to be larger the longer the power outage. Retail marketers may also demand penalty clauses in their access agreements with distribution utilities for failure to restore distribution service within a reasonable amount of time after the event that took out the lines. Incentives to restore power after an outage could be even greater under competition if the utility remained vertically integrated into energy sales. The negative publicity associated with a sustained outage could result in the utility losing retail energy customers to competitive suppliers.

Bulk Power Reliability

Unlike the distribution grid, the security of the bulk power transmission grid is a classic example of what economists refer to as a *public good*. If a utility makes an investment or curtails a power transaction to maintain system security, not only will that utility and its customers benefit from that action, but all the utilities and customers attached to the grid will also benefit. Because all the benefits of promoting system security cannot be captured by the utility making the investment, if left to their own devices, utilities will tend to spend too little on security-enhancing activities and seek a free ride on the efforts of others.

This problem of *free riding* is less of an issue if utilities do not compete with one another and are subject to rate-of-return regulation (see Chapter 8). In a world with interconnected, vertically integrated and regulated electric power monopolies, utilities do not have to worry about how investments they make to improve grid security might benefit their competitors because they have no competitors. Moreover, when costs are fully recoverable through regulated rates, the utility has no incentive to skimp on expenditures to support system security.

As markets become more competitive, however, utilities are likely to be much less willing to make expenditures voluntarily that will benefit the larger community of grid users, some of which are their direct competitors. Some fear that under competition utilities will reduce grid-related maintenance expenditures and seek to push the costs of maintaining system security off onto others. Competition also creates greater incentives to reduce costs, which make utilities even less willing to make nonessential or nonmandated expenditures for which their customers are not willing to pay.

Competition has already begun to increase the number of participants and the volume of transactions in wholesale electricity markets, with associated increases in power shipments. Even greater increases are expected as more retail markets open to competition. This increased volume of power transactions and shipments brings with it an increase in the probability that some component of the transmission grid will fail. This higher probability of failure, in turn, creates an enhanced need for system security measures.

Increased participation in electricity markets by a wide variety of players—ranging from power marketers and brokers to independent "merchant" generators—further complicates the problem of assigning the responsibility to maintain reliability. Traditionally, in most NERC reliability regions, the quantity of spinning reserves (see Chapters 2 and 10) and other reliability-related or ancillary services necessary to maintain system security were centrally determined and procured by PCA operators, and the costs of those services were averaged over all system users. But in a competitive world, some market participants may desire different charges based on each user's contribution to system reliability. For example, some would argue that utilities or independents that have longer or more frequent unexpected outages of their generators should pay a greater share of the cost of reserve generation than other system users.

Ensuring Reliability—or Stifling Competition?

To maintain the security of the bulk power transmission grid, PCA operators and security coordinators may need to interfere with the commercial transactions on the electricity grid. In particular, if some component of the system is approaching an emergency situation—one in which it would not be able to pick up its share of the altered power flows in case of a contingency—then the operator will need to curtail some commercial transactions. These types of curtailments to prevent system failures have come to be called *transmission loading relief* procedures (TLRs). Ideally, TLRs are undertaken in a nondiscriminatory way, with those most willing to bear the interruptions being curtailed before those that have higher costs of interruption. Realistically, the system operator may not have the information necessary to make this type of judgment. In such regions as California, the ISO is often precluded from using economic information to make decisions about which transactions to curtail.

In some instances, power marketers and other market participants have accused transmission system operators of implementing TLRs in somewhat arbitrary ways that impede the proper functioning of electricity markets. Transmission operators have been accused of being overly conservative in their estimates of how much power flow the system can handle and thereby unnecessarily limiting transactions and raising electricity prices. Transmission operators have also been accused of curtailing transactions in an arbitrary fashion that does not allow the most valuable transactions to proceed and curtail those of lesser value.

Electricity system operation is not an exact science, and thus system operators need some discretion about whether or not transactions will be interrupted to maintain security, and sometimes about which transactions to interrupt. The fact that discretion and expert judgment are required means that it may be difficult to distinguish a reduction in permitted energy flows taken to protect system security from one taken for other, perhaps anticompetitive, reasons. Making this distinction can be particularly problematic when the system operator is aligned with a particular player or group of players in the generation markets.

Better developed markets for use of the transmission system might help to reduce the opportunities and incentives for transmission operators to arbitrarily curtail the use of the transmission grid to forestall an emergency situation. In Order 2000, the Federal Energy Regulatory Commission (FERC) calls on RTOs to manage congestion in a way that promotes the efficient production of electricity and allows for a secondary market for transmission rights where holders of the rights to transmit power can trade those rights with others. If transmission rights are tradable, markets can help to allocate the use of scarce transmission resources to those who value them most. Transmission operators can use prices from these markets to help guide their decisions about which transactions to curtail when curtailment is deemed necessary and when to expand capacity in the future.

Ensuring Reliability in a Restructured Market

Given the potential threats to reliability posed by electricity restructuring, legislators and energy regulators should develop a strategy to protect system reliability as they design and implement policies that set the course for restructuring. Policymakers need to address both the adequacy and the security components of the reliability issue.

Generation Adequacy

In a competitive market, adequacy of the electricity supply will be resolved largely in the marketplace. During periods of peak demand, when available generating capacity is more fully utilized, the market price of electricity is bid up. This creates two incentives. First, it creates an incentive for those electricity consumers who can interrupt or postpone their electricity usage to do so. Second, it creates an incentive for electricity generators to build more peaking capacity to take advantage of the potential profits associated with high peak-period prices. These two activities should help to reduce both the chance of electricity shortages and electricity price spikes during peak periods.

For the market to correct supply inadequacy problems, consumers must be active participants. Those consumers who are willing and able to curtail their electricity usage when prices rise need to be able to see a time-varying price that would provide them with a signal indicating when the market price of electricity exceeds the value they place on consuming that electricity. Several of the early competitive electricity spot markets, including England and Wales and California, were originally designed largely as a forum where suppliers compete to meet a fixed level of demand; electricity consumers were not active participants in the markets. Allowing consumers to bid a reservation price—above which they would be willing to curtail or interrupt their energy use—would help to prevent large price spikes and would help to reveal the true market value of an adequate power supply.

During the transition to a competitive electricity market and perhaps beyond, there will likely still be centralized monitoring of power supply adequacy by NERC or its successor organization (see below). PCA operators and others will also continue to try to ensure that there is sufficient capacity to meet anticipated demand.

Transmission services will continue to be regulated after restructuring. The transmission regulator, primarily FERC, will be responsible for setting prices for transmission service and for overseeing investment. Because a monopoly transmission provider could have an incentive to restrict its output, the regulator will need to take an active role in ensuring that there is adequate transmission capacity to meet consumer needs. (For more on transmission regulation, see Chapter 8.)

Security and New Reliability Institutions

Institutions developed to promote system security in a world where electricity supply was dominated by vertically integrated electric monopolies will not work well in a more competitive environment, for at least two reasons. First, complete participation in the governance of these organizations has historically been limited to traditional utilities; participation by independent power producers, power marketers, and other nonutility entities, all of whose ranks continue to grow with the introduction of competition, generally has been either limited or precluded. For these institutions to be impartial and to be trusted by all market participants, they are being expanded to include broader representation on their governing boards. Second, with the erosion of incentives for voluntary cooperation, reliability institutions need to have greater authority to enforce the rules for maintaining system security.

To address both of these concerns, the NERC Electric Reliability Panel, a committee created to recommend new ways to ensure reliability in a competitive marketplace, recommended in 1997 that NERC be transformed into a new organization called the North American Electric Reliability Organization (NAERO). This transformation is currently under way. NAERO will have a broader membership base and governance structure that is representative of all players in the electric power market and the public at large, and it will receive funding from a wide range of sources. The panel also recommended that NAERO be officially recognized by the federal government as a self-regulating organization (like the National Association of Securities Dealers) that has authority to enforce its rules and to collect funds to support its operation. Several federal bills have been introduced that would give NAERO and its regions statutory authority to enforce compliance with its reliability standards and rules among all market participants. Most of these bills would establish FERC as NAERO's federal overseer. FERC would approve NAERO policies, hear complaints against it, and back up its enforcement decisions.

An important challenge for NAERO will be to devise rules for determining the pattern of system redundancies that are compatible with reliability goals and with the efficient operation of competitive wholesale markets. With more decentralized planning of generation investment, NAERO will also need to develop new ways to evaluate the trade-offs between investment in generation and transmission. If customer participation in electricity spot markets increases, NAERO may need to broaden its definition of service reliability from an engineering goal to an economic concept that explicitly incorporates the trade-offs between price and continuity of service.

Another institutional change given a boost by the issuance of FERC Order 2000 in December 1999 is the widespread development of independent regional transmission organizations, either in the form of an ISO or an independent transco that both owns and operates transmission facilities (see the box, "FERC Orders," in Chapter 3). Assuming that they are truly independent, RTOs should eliminate the incentives that vertically integrated transmission operators might have to curtail transactions in the name of system security when unnecessary or to restore system security by curtailing transactions of competitors while allowing their own facilities to continue to operate. True independence should help to create more confidence that transmission operators are motivated purely by reliability concerns. FERC's call in July 2001 for the creation of five large RTOs encompassing the continental United States has brought about suggestions that these RTOs might replace NAERO regions as overseers of regional reliability.

Security and Incentives

Another policy tool that can be used to promote system security in a restructured industry is the creation of incentives to use the transmission and distribution grids more efficiently. On the transmission side, the use of some form of congestion pricing of transmission services could provide incentives for generators to locate in uncongested regions. Depending on who benefits from the collection of congestion payments, such pricing might also lead to expanding the transmission grid in a way that alleviates bottlenecks (see Chapter 8).

Reliability of both the transmission and distribution parts of the system could be enhanced by greater penetration of distributed energy resources. This category of resources includes small-scale generators—such as gas- or diesel-fired microturbines; fuel cells; or micro-windmills and local power-storage facilities, such as flywheels, batteries, or pumped-storage units. Distributed resources are located close to loads and thus reduce reliance on the transmission and distribution grids. This reduced usage in turn will decrease the probability of outages on the grids.

Currently, distributed resources rarely can compete with the cost of central station generation. Some distributed power technologies may be economical for customers who place a particularly high value on reliability or who are in areas where a heavily used distribution system would be costly to upgrade. Regulations that set uniform average-cost prices for transmission and distribution, however, will fail to indicate particular locations where installing distributed generators might be worth the cost. In addition, better monitoring, communications, and control capabilities are needd to provide information on how using these technologies affects system performance and to adjust their use in response to overall system conditions. A lack of established standards for interconnecting small generating units with the grid also may discourage their deployment by those not affiliated with the local utility.

CHAPTER

12

State and Federal Roles

State governments have been the key actors thus far in developing and implementing policies to encourage retail electricity competition. A policy question has been whether states are acting quickly enough, or whether the federal government should step in to encourage or force them to open markets by a particular time.

Keeping control with the states allows the nation as a whole to learn from what worked in one place and what did not work so well someplace else. One size may not fit all, in that the benefits of opening markets may be considerably greater in some states than others. In addition, to impose a federal solution could cause needless and costly difficulties in trying to amend or reverse the delicately balanced solutions achieved by states that are moving ahead in opening retail markets.

However, a presumption that state actions might reflect a proper balance of interests is less convincing when that state's decisions have effects that go across their boundaries. When interstate effects are significant, the federal government can help to improve policies by serving as a venue where all affected parties have a say. Specific areas in which the federal government can play an effective role include reforming existing federal laws that may inhibit competition, regulating interstate transmission grid prices and operation, enhancing market liquidity, enforcing antitrust and environmental laws, and coordinating commercial standards and practices. Also, states themselves may be able to negotiate solutions and set up regional authorities to manage issues that affect an entire region but not the nation as a whole.

To the extent that average persons think about regulation of major industries such as electricity or telecommunications, they probably assume that the regulator is in Washington, DC. Such a predisposition would not be surprising. National debates and decisions by Congress, federal regulatory agencies such as the Federal Communications Commission and the Federal Energy Regulatory Commission (FERC), and the courts grab the headlines and get national media coverage.

But in practice, state public utility commissions carry much of the responsibility for regulating utilities, including virtually all of the prices consumers pay (see Chapter 3). There is an extensive federal role, to be sure. FERC has the authority to regulate the prices paid for electricity at the wholesale level, which we may think of as sales to distribution companies and any other entity that resells electricity, particularly at retail to the households, offices, shops, and factories that actually use it. FERC regulates the prices, terms, and conditions for the interstate transmission of electricity, including the operating rules for independent system operators and regional transmission operators. (Chapter 9 discusses the roles of FERC, the Department of Justice, and other federal agencies regarding mergers.)

Yet state governments, legislators, and public service commissions retain the authority to set the rules and rates for retail sales of electricity. Included in these rules are whether, when, and how to implement retail competition. As was discussed in Chapter 4, the first states to adopt retail competition—California, Massachusetts, New York, Pennsylvania, and Rhode Island—were among those with the highest electricity prices in the country. Virtually all states have taken some legislative or regulatory action to at least look into the possibility of opening electricity markets to retail competition. As of mid-2001, nearly half the states plus the District of Columbia had formally adopted plans to institute retail competition by some foreseeable date in the next few years.

The state-driven aspect of the move to open retail electricity markets raises first the matter of trying to help consumers, businesses, regulators, and elected officials across the country become aware of the many complexities associated with increasing competition. But it also has raised and continues to raise the question of whether the federal government should do more to promote competition in electricity, efficient transmission, and control of environmental effects.

Positions on Restructuring Authority

A first take on the different views regarding the extent of the federal role in promoting retail electricity competition is to place them along a simple continuum, from "do nothing" to "do it all." This is something of an oversimplification. As we will see below, federal actions can take place along numerous dimensions. Nevertheless, to get a handle on the "federal versus state" question, it helps to begin with a look at the broad policy options.

At one end of the continuum, Congress and federal regulators could let the states continue to proceed on their own. As was noted above, many states are exercising the options available to them and instituting retail competition now or in the next few years. If retail competition is the goal, a fair question may be whether a federal role is fundamentally necessary.

At the other end of the continuum would be a federal mandate. Some states are moving quickly to choose and implement retail competition. Others are reluctant to move ahead at all, and some have reversed course. In prior sessions of Congress, numerous bills have been introduced that would mandate retail competition in all states by a particular date. Many of these bills also include provisions regarding competition policy issues, environmental concerns, and reforms of current federal statutes affecting the electricity industry.

A third, middle way between these two alternatives would be to require states to institute retail electricity competition by a particular date, but allow them to "opt out" if, after some process yet to be specified, its government formally decided not to go that route. A proposal of this nature was offered by President Bill Clinton's administration in 1999. Such a plan would leave discretion to the states, but would force them to consider explicitly whether or not to adopt retail electricity competition. In this view, retail competition ought not fail simply out of neglect or a lack of political impetus to put the issue on that state's policy agenda.

Retail Deregulation and the States

Coming up with the best policy in complex regulatory and deregulatory matters requires considerable understanding of how the relevant industrial technologies operate and how the relevant economic markets behave. Electricity restructuring is rife with issues of this sort, where legislators and regulators have to weigh issues of cost, reliability, competition, environmental protection, and equity in designing restructuring policies. Acquiring the expertise appropriate to make good judgments regarding each of these issues, much less how they might be balanced, is a nontrivial task, to say the least.

In economic jargon, this expertise requirement can create an economy of scale in making and implementing public policy. Having each state develop this expertise on its own can lead to expensive, redundant efforts in hiring and developing the knowledge and skills to address these many issues. Accordingly, just as most of us take our cars to a mechanic rather than fix them ourselves, states might elect to let the federal government take on this effort on behalf of all the states. The observed result could be a federal presence in electricity (or other) policy issues, even when the stakes are primarily local.

There is, however, a flip side to this coin. When matters are sufficiently complex and uncertain, the ability to learn becomes a particular virtue. To the extent that states have the ability to design appropriate retail competition policies, letting them do so creates dozens of laboratories where different market designs, industry restructuring methods, and reliability policies can be tested. States can learn from each other, and the federal government can learn from them. Much of the impetus to deregulate airlines came from the opportunity to observe the successes in California and Texas, where substantial in-state airline markets had been opened to competitive entry and pricing. A risk in letting the federal government preempt policy is that it might attenuate this learning process, perhaps leading to rules that might turn out to be inferior.

A second risk of federal preemption is that "one size" may not "fit all." The premise that there might be scale economies in expertise warranting federal preemption of state retail competition decisions is that the policies developed for state A will also be the best for state B. This may but need not be the case. States differ greatly in population density, industrial base, climate, regulatory history, and generation portfolio. All of these may influence which policies would best implement retail competition, how fast it should move, and who deserves compensation from whom. Federal mandates for retail competition could involve the national government in managing state markets in the face of these hard local political and economic questions.

As happens so often, one is left with having to balance benefits against costs. Here, the benefit of federal intervention is economizing on the expertise necessary to best implement retail competition. The costs involve a potential loss in the opportunity to learn from different state laboratories and the risk of imposing a uniform solution to problems that vary across the country. The record suggests that this tilt may be against federal intervention. States are leading the charge to bring retail competition to their constituents, suggesting that the scale economies in expertise are not compelling. They have not waited for a nominally more expert federal government to tell them what to do.

Moreover, the states are doing things quite differently in terms of timing, market design, and separation of generation from transmission and distribution. Perhaps most important, they are reflecting their different legal, historical, and political environments in distributing among interest groups the gains from competition. The extant efforts of states to develop retail electricity markets, and the variation as to how and when they do so, is itself reflected in federal legislative proposals that would "grandfather" these existing arrangements. To some degree, the willingness of many states to take giant steps toward opening retail electricity markets had preempted much of the role the federal government might take in prescribing how such markets should function.

Federal Intervention

In practice, decisions regarding the division of power between the states and the federal government in expanding electric competition to the retail level are, virtually by definition, quintessentially political. Statutory history regarding how authority over electricity has fallen to the states and the federal government, and larger constitutional traditions regarding the right of Congress to make legislation affecting "interstate commerce," will also be major influences on the outcome of this debate. But the inevitable influence of politics and law on these questions need not preempt thinking about the allocation of responsibility as a policy issue itself.

The considerations listed above—the ability to learn from what different states do, the strong possibility that one size may not fit all, and the fact that many states have already acted—argue in favor of leaving the decisions to the states. Moreover, we might also add that a presumption in favor of keeping decisions as local as possible is consistent with the fundamental premise underlying restructuring, that individual buyers and sellers are better than central regulators at making decisions about prices and purchases (see box on next page).

However, that presumption need not hold when the actions taken in one state affect those in others. This can happen in two important ways. The first involves insufficient competition and market power (discussed in more detail in Chapters 7 and 9). If suppliers of a commodity in one state have cornered the market in a particular good desired by consumers in other states, those firms and their state will have an incentive to hold supplies off the market, raising prices in other states to increase profits.

A second important interstate effect involves costs that lie outside the market, in the sense that those who bear the costs of others' choices cannot either charge those others for the costs they impose or pay them to act differently. The standard term for

Should the Quality of Government Matter?

Any public policy decision, regarding retail electricity competition or anything else, should be responsive to the wishes of the constituents of the government body making the decision. Without interstate effects, a state's elected legislators and elected or appointed regulators are closer and more likely to be responsive to its constituents than are more distant federal representatives and policy practitioners in Washington.

At its heart, this is but an extension of the principle underlying expanding the role of market forces in the electricity industry. The fundamental justification for leaving decisions to the market is that buyers and sellers are the best judges of their own welfare. By analogy, if consumers should make decisions that affect only themselves, so too should states.

But consumers and states are not the same. As philosophers, political scientists, and economists have pointed out for centuries, governments may not faithfully reflect and properly balance the interests and values of their constituents as consumers, businesses, environmentalists, and citizens. Relatively small groups with large stakes in the outcome may be able to influence government decisions out of proportion to the weight that they might have in an ideal cost–benefit test or democratic process.

But are state governments more vulnerable than the federal government to undue influence from particular interests? "Inside the Beltway," this seems a widely held if rarely stated supposition. Yet the argument could go the other way. Vesting authority with the federal government may enable groups with organizational advantages to achieve their ends in a single step rather than forcing them to exert their influence over numerous state and local governments.

Responsiveness should not be judged simply on the basis of the decisions that get made. The public in state A may be generally inclined to promote economic efficiency and rely on markets and, thus, to adopt retail electricity competition. Citizens in state B may believe that regulation is a better way to obtain electricity and perhaps promote some noneconomic goals. One ought not automatically infer that those who profit from retail competition have "captured" policymakers in state A, or that those who profit from regulation have similarly distorted policies in state B.

Because it is so hard to tell in advance whether local or national governments are more faithful to their constituents, claims that one level of government is better than another probably ought not carry much weight. In deciding whether retail electricity competition should be a matter for the federal government or for the states, the magnitude of interstate effects, costs of developing expertise, and value of running numerous policy experiments in the various states should be more important factors.

such nonmarket effects is *externalities*. The textbook example of these externalities is air pollution, where the absence of a market for environmental rights means that polluters will treat air, water, or land as if it is a free waste dump. If a manufacturer's pollution stays within a state line, the state government has the ability to decide when the benefits of cleaner air are worth the cost of reducing pollution. But if the pollution crosses a state line, the state will have little incentive to factor harms to those outside its boundaries in deciding what kind of environmental controls to place on its factories. We discuss these environmental issues in Chapter 15.

Important legal doctrines do not entirely match the substance of this idea. In practice, the "commerce" clause in Article I, Section 8, of the U.S. Constitution granting Congress the power "to regulate commerce . . . among the several states," typically requires only an interstate flow of labor, capital, goods, or services. The commerce clause can be satisfied by the mere fact that some aspects of the enterprise cross state lines, but need not include demonstrable benefit or harm that would warrant policy intervention. And conversely, the presence of interstate harm need not ensure a legal role for the federal government. In antitrust, the "state action" doctrine allows states to impose "clearly articulated and actively supervised" policies that limit competition, even if the effect of that competition is to raise prices outside the state.

Of course, noneconomic considerations can matter. The nation as a whole may want to interfere with an otherwise well-functioning market to correct for inequalities in wealth, welfare, opportunities, or legal rights. Some of these inequalities may be concentrated in particular states, implying that national efforts may involve some interference in state or local decisions. These may involve protecting certain suppliers from competition, controlling prices, subsidizing the purchase of particular goods and services, and otherwise regulating markets as a way to address social concerns.

What about Electricity?

If citizens of state A perceive that the retail competition is worth instituting, or if they perceive otherwise, the story so far suggests that the decision should be their prerogative. The obvious benefits and costs of opening electricity markets in a given state fall within the boundaries of that state. To some extent, a decision not to open a market to competition may deny a profit opportunity to electricity providers outside the state. In general, however, if power markets are competitive and cover a broad geographic area, power sellers can make similar profits in other markets. Moreover, buyers have no duty to buy from one seller rather than another just to help the former seller make more money.

But that prerogative ends as long as the benefits and costs fall within the state. Not all aspects of retail electricity competition come out equally when examined in terms of interstate effects, expertise, and experimentation. Even if one were to decide that decisions on whether and how to implement the opening of retail electricity markets should be left to the states, some areas remain in which a federal role is likely to be necessary.

Reforming Federal Laws

During their history as generally regulated monopolies, electric power utilities have long been subject to federal statutes that affect their corporate organization and operations. Reforming or repealing these statutes may be useful in promoting the efficient development of retail markets. Two examples (presented in Chapter 3) stand out. Provisions in the 1935 Public Utility Holding Company Act could limit the ability of a utility in one region of the country to sell power in another region, possibly limiting competition and the development of useful "brand names" in electricity, although utilities are able to get around these restrictions by creating subsidiaries that qualify as exempt wholesale generators under the Electricity Policy Act of 1992. The 1978 Public Utility Regulatory Policies Act imposes obligations on utilities to purchase electricity from power producers who use renewable fuels. As entrants do not face these obligations, Public Utility Regulatory Policies Act (PURPA) policies could lead entrants to supply power when an incumbent would be able to do so at a lower cost. Most observers of the electricity sector believe that, at a minimum, Congress should amend federal law to remove these impediments to competition.

Transmission Grid Operation

Although the different portions of transmission grids in different states may be owned by different companies, the fact that electric currents take all paths between destinations implies that the "grid" is, in effect, a single entity that crosses state lines (Chapter 2). The inherently interstate character of transmission justifies the traditional role the federal government has had in regulating transmission and wholesale power markets. Through its orders regarding open access to the transmission grid and the formation of regional transmission organizations, FERC continues to oversee this inherently interstate complement to open retail electricity markets. FERC's recent advocacy of creating four regional transmission organizations that would serve respectively the Northeastern, Southeastern, Midwestern, and Western United States, as discussed in Chapter 8, exemplifies this federal role.

Other regulatory considerations may also create interstate effects. Chapter 8 discussed the effects of pancaking (i.e., forcing generators to pay separate fees to each transmission system in each state between it and its customers). For an additional illustration of pricing issues, suppose that the costs of an interstate transmission grid are averaged over all users when computing transmission rates. Suppose that consumers in state A want more electricity. Two options for meeting this demand would be to build a generator in state A or to build a transmission line between states A and B. If the costs of the latter are spread across states A and B, consumers in state B are subsidizing the transmission choices of state A. Hence, state A may opt for additional transmission to deliver power from other states, because those in state B share the costs, when building generation in state A might be a less expensive means for providing the added power. An interstate authority with power over transmission siting decisions might be necessary to promote a more efficient electricity delivery infrastructure.

Market Liquidity

Electricity markets, whether wholesale or retail, may cross state lines, depending on the prices and capacities of the transmission systems that carry electricity from one point to another (Chapter 9). In and of itself, the multistate nature of an electricity market need not warrant a federal role in determining whether a particular state should decide to participate in those markets. However, in parts of the country where those markets are particularly "thin" (i.e., have few participants), a decision by one state to prevent customers from choosing their own electricity supplier—or to design its retail or wholesale markets in an ineffective manner (see box on next page)—could hamper the ability of electricity markets to meet the demands of consumers in other states. If one state's decisions on whether and how to implement opening its electricity markets affect electricity prices and availability in other states, the federal government may be where those implementation decisions should be made.

"Retail Choice" and Wholesale Markets

As jurisdiction between wholesale and retail electricity markets is divided, it is up to a state government to decide whether to open retail electricity markets (i.e., to allow the residents and businesses within it to choose their own electricity supplier). If a state opts for retail choice, however, it creates the potential for new electricity marketers, and thus the need for these new entrants to obtain electricity in wholesale markets. In addition, some states have decided that to promote competition in electricity markets, they have required distribution companies to spin off their generators.

The presence of new retail electricity providers and the need for distribution utilities to purchase electricity implies a need for a wholesale market in that state in which these companies can obtain the electricity that they sell to their consumers. Consequently, in the absence of existing wholesale markets, states that opt for retail choice also end up planning the wholesale markets that go with them. In California, for example, the institutions at the core of its endeavor to open electricity markets—the California Independent System Operator and Power Exchange, as described in Chapters 4 and 5—were wholesale market institutions.

Decisions by states to create wholesale power markets to accompany the opening of retail markets do not indicate a shift in jurisdiction. As with other wholesale markets, those created by states require approval by FERC, in terms of how well they promote the objectives of promoting competition and reliability as set out in FERC Orders 888, 889, and 2000. Such approval, however, neither mandates "one size that fits all" nor ensures that wholesale markets will perform effectively, as the contrast between California and Pennsylvania–New Jersey–Maryland (Chapter 4) has shown.

Antitrust

A core policy concern is ensuring that electricity markets perform competitively. The antitrust laws promote competition by encouraging independent supply decisions and preventing collusion among sellers. Depending on transmission capacity and rates, however, the prices consumers pay for electricity in state A may depend on how intensively generators in states B and C are competing with each other. Prices in state A may also depend on whether the transmission lines that run from state D to state A offer genuinely nondiscriminatory access at reasonable rates.

Federal antitrust enforcement and other policies will likely be necessary to promote competition. These policies would ensure independence of generation from regulated transmission (Chapter 7). These agencies may also have a role in preventing anticompetitive mergers that would limit competition. Also, additional authority may be necessary to promote the divestiture of generation assets, to increase the number of significant independent power producers in retail electricity markets (Chapter 9).

Environmental Regulation

Electricity production generally comes with environmental problems, chiefly air pollution associated with burning fossil fuels (discussed in Chapter 15). Some of these pollutants may travel long distances in the atmosphere before adversely affecting the environment. Accordingly, emissions from generators in state A may impose pollution-related harms on those who live in states B and C. In the case of carbon dioxide and greenhouse gas emissions, the effects may not simply be interstate, but global. When the effects of the pollution fall outside the jurisdiction of the state in which a polluter sits, a federal role, most likely through the Environmental Protection Agency, may be necessary to ensure that the polluter bears the environmental costs it imposes on those in other states.

Coordinated Standards and Practices

As electricity markets develop, the "rules of the game" regarding terms and conditions of sale, contracts, implied warranties, product liability, and the full breadth of procedures and practices for doing business will have to be settled. To some extent, competition among the states to come up with best commercial practices could redound to the benefit of all, just as there can be broad national benefits from allowing states to determine how to make and implement choices regarding the design of retail electricity markets.

Some observers of the electricity industry suggest that uniform commercial standards and practices can reduce everyone's cost of doing business. Establishing such standards and practices may require action at the national level, perhaps through congressional legislation or rules implemented by the Department of Energy or FERC. Negotiations among market participants may be best, although public oversight may be useful in preventing such negotiations from reducing competition.

Supporting Social Programs

Last but not least, electricity markets have been subject to a variety of social benefits programs (discussed in Chapter 16). These include conservation programs, subsidizing research, development, use of renewable power sources, and policies to ensure the availability of electricity to low-income households and rural areas. The justifications for these programs will frequently be national in character. For example, research and development into renewable power sources would make such power available across the country, and not just in the state that funded the research. The social interest in ensuring widespread access to and use of electric power is also likely to cross state lines. Accordingly, the federal government, through the Agriculture, Commerce, and Energy departments, and perhaps others, may retain a continuing interest in these programs.

Regional Authorities

Finally, it is important to note that "interstate" and "national" need not be the same. A group of states with a particular common problem may be able to institute a regional authority with the ability to make decisions that adequately balance benefits and costs among electricity producers, transmission and distribution companies, and customers in the entire area. One might think of these as public analogues to interstate power pools, reliability councils, and regional transmission organizations. Examples of such arrangements in other contexts include the Port Authority of New York and New Jersey and the Chesapeake Bay ecosystem restoration program established by the District of Columbia, Maryland, Pennsylvania, and Virginia (along with the federal Environmental Protection Agency). As retail electricity markets evolve, we may well see states come together in formal and informal ways to resolve interstate problems, without necessarily requiring federal legislation or regulation.

CHAPTER

13

Public Power's Role after Restructuring

Unlike most of the other industries that have made the transition from regulation to competition, the electricity sector has a substantial nonprofit component. Roughly 25% of all retail electricity sales in the United States come from publicly or cooperatively owned utilities. The combination of privately and publicly owned utilities (at local, state, and federal levels of government) operating under different objectives and rules greatly complicates the task of restructuring the electric power industry. Together these utilities are commonly referred to as the public power providers.

If public power providers are going to be allowed to compete with privately owned generators, policymakers should address the distinguishing characteristics of public power. For competition to function well, all competitors should have the same opportunity to compete. Special privileges granted to public power providers—such as the ability to issue tax-exempt debt and preferential access to low-cost hydroelectric power from federally owned facilities—gives them an advantage over their competitors. In addition, both the unique regulatory status of public power providers and rules that limit their ability to participate in regional transmission organizations could inhibit the development of geographically large wholesale electricity markets. Many federally owned hydroelectric generation facilities have multiple purposes, such as flood or navigational control, that need to be recognized in debates over the role of these facilities in competitive markets.

How public power will evolve in this era of competition remains an open question to be decided at different levels of government. The federal government will be responsible for redefining the roles of federal power marketing authorities and the Tennessee Valley Authority. Decisions about whether municipal utilities or rural cooperatives should continue to hold an exclusive franchise for retail electricity sales are best made at the local level.

Restructuring the electric power industry and bringing competition to generation and retail sales markets would be complicated enough if the industry were populated exclusively by for-profit investor-owned utilities. The task becomes even more difficult when the roughly 2,900 publicly or cooperatively owned utilities are added to the mix. Taken together, publicly and cooperatively owned utilities were responsible for roughly 25% of the nearly 3,240 billion kilowatt-hours of electricity sold by utilities to businesses and households in 1999. These publicly and cooperatively owned utilities tend to face different forms of regulation from investor-owned utilities. They also are granted special privileges not extended to investor-owned utilities. This special treatment could provide publicly owned and cooperatively owned utilities with advantages over their investor-owned competitors and could keep electricity markets from operating as efficiently as possible.

Electricity is not the only industry in which we have both privately owned providers competing with nonprofits or publicly owned entities. Private schools compete with public schools for students at all levels, from elementary grades through graduate and professional school. Publicly owned and nonprofit hospitals compete with privately owned and for-profit hospitals in hiring doctors and nurses and in serving patients.

In these industries, the most prominent concerns regarding different treatment have been subsidies and the applicability of the antitrust laws. The subsidy question is more notable in education—the political debate over vouchers is, in some way, about whether private schools should get tax support similar to that for public schools. With regard to antitrust, the nonprofit or publicly owned standing of universities and hospitals has been invoked as a reason to exempt them from laws prohibiting alleged agreements to fix prices (e.g., scholarship grants) or to permit perhaps anticompetitive mergers.

The debate over how to bring competition to electricity generation and retail sales markets has highlighted several differences between publicly and cooperatively owned utilities and investor-owned utilities. These differences can be categorized into three types: financial, regulatory, and scope. The first category refers to the special privileges granted to public utilities and cooperatives, which include preferential access to low-cost hydroelectric power produced at federally owned facilities, the ability to issue tax-exempt debt, and exemption from income tax payments. The second category refers to the exclusion of publicly owned and cooperatively owned utilities from the state and federal regulatory structures governing investor-owned utilities and other rules restricting the private use of capital funded with tax-free debt. The third category refers to the fact that many federally owned hydroelectric generation facilities have multiple purposes, such as flood or navigational control, in addition to electricity production.

This chapter provides an overview of how the roles of public power providers, cooperatively owned utilities, and federal utilities and federal power marketers might change as a result of electricity restructuring. We define what is meant by each of these terms and discuss the role of publicly owned utilities in the push for wholesale competition. Then we consider whether a so-called level playing field exists for public and private utilities and how the roles for publicly owned and cooperatively owned utilities are evolving. We next examine the ways that the pres-

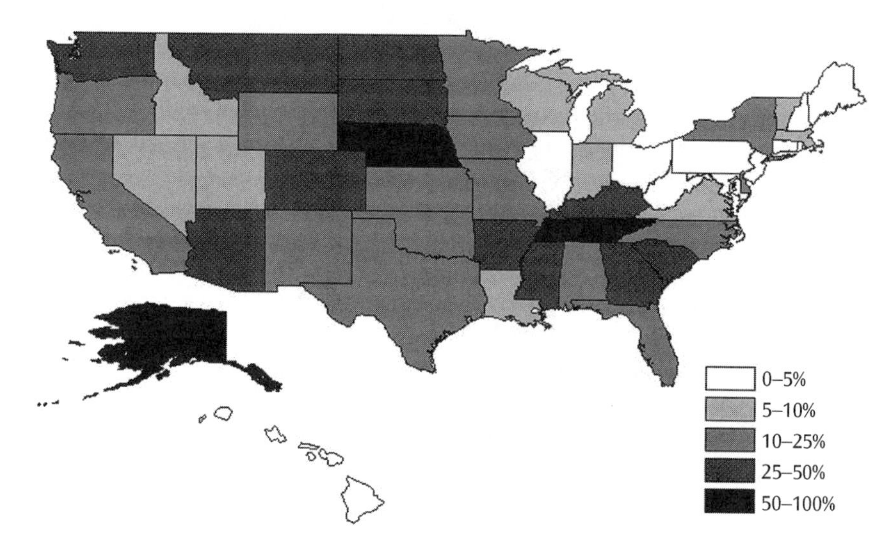

Figure 13-1

Percentage of U.S. Electricity Customers Served by Cooperative or Publicly Owned Utilities, 1999

Source: U.S. Department of Energy, Energy Information Administration, *Electricity Sales and Revenue, 1999*, "Sales to Ultimate Consumers by Class of Ownership, Census Division and State, 1999," Table 9. http://www.eia.doe.gov/cneaf/electricity/esr/t09.txt (accessed January 20, 2002).

ence of federal power is affecting restructuring and some policies that have been proposed to redefine the role of federal power in a competitive market. We conclude with some thoughts about which jurisdictions should be responsible for various types of policy decisions.

What Is Public Power?

Technically, the term *public power* refers only to those electric power suppliers that are owned by the government, including municipal utilities, state-owned utilities, and federal power-marketing agencies. However, utilities that are owned cooperatively by their customers instead of by the government have much in common with public power, so we include them here. The Tennessee Valley Authority, a federal corporation that generates electricity in addition to helping economic development and watershed management in the Tennessee Valley, is also included in most discussions of the implications of electricity restructuring for public power. The map in Figure 13-1 shows the percentage of customers in each state currently served by publicly owned or cooperatively owned utilities, the two categories of public power that serve retail customers directly.

Public power has existed since the last two decades of the 1800s, when the power industry was in its infancy. At that time, public power systems spread to rural areas where electrification was not profitable due to low population density. The trend continued with the growth of the nation and a demand for more electricity. Federal electricity production began during the Great Depression in the 1930s as a stimulus for economic development in rural and disadvantaged areas. The following three subsections provide additional detail on publicly owned and other nonprofit utilities.

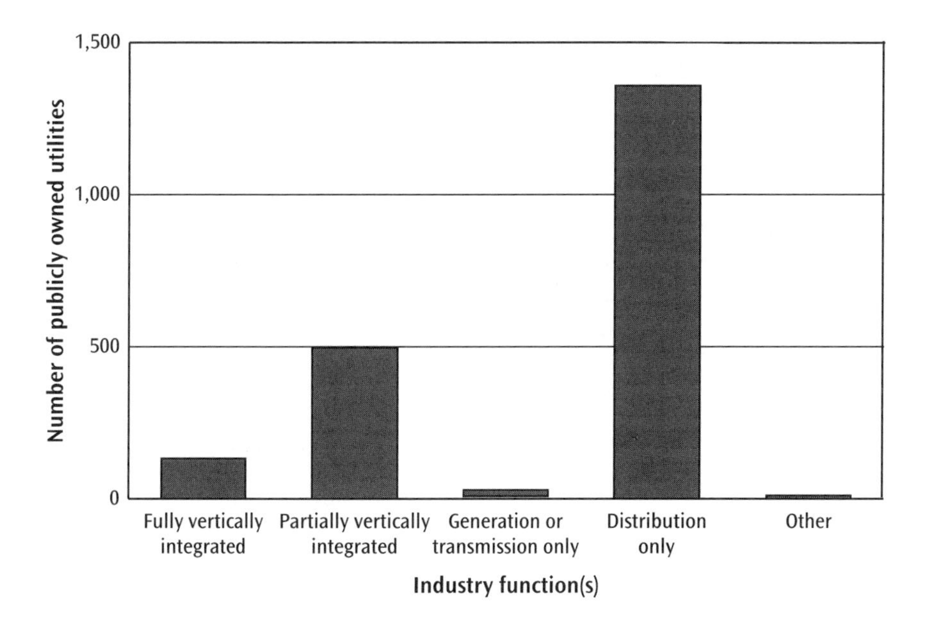

Figure 13-2

Industry Functions Performed by Publicly Owned Utilities, 1998

Note: Fully vertically integrated, publicly owned utilities perform generation, transmission, and distribution. Partially vertically integrated, publicly owned utilities perform two of three industry functions. The category "other" includes utilities that perform equipment maintenance and purchasing services for their parent utility.

Source: U.S. Department of Energy, Energy Information Administration, *The Changing Structure of the Electric Power Industry 1999: Mergers and Other Corporate Combinations* (Washington, DC: U.S. Department of Energy, 1999).

Publicly Owned Utilities

Municipal utilities represent the largest group of publicly owned utilities (POUs). Most municipal systems operate only as electricity distributors for small communities, purchasing most or all of their power requirements from federal and state agencies when possible and from investor-owned utilities when necessary. However, some municipal systems are vertically integrated into generation and transmission functions. As Figure 13-2 indicates, however, few POUs are fully vertically integrated, and most provide only distribution services, including retail sales. Municipal utilities serve some of the nation's largest cities, including Jacksonville, Los Angeles, San Antonio, and Seattle. In addition to providing electricity, some municipal utilities may provide a combination of natural gas, water, wastewater, and Internet access services.

Public power districts, irrigation districts, and state authorities are POUs that serve a broader geographic area and that tend to provide a different set of public services in comparison with municipal utilities. Public power districts are most common in the West Coast states and Nebraska. At the state level, irrigation districts usually provide regional water-related services and generally focus on generation and transmission of electricity. For example, the Merced Irrigation District in California and the Lower Colorado River Authority in Texas both generate and transmit power for other utilities and also provide agricultural water service and maintain recreational areas. State authorities, such as Santee Cooper in South Carolina and the New York Power Authority, are state-owned utilities that mainly provide electricity directly to municipal utilities and cooperatives. Unlike the New York Power Authority, about 125,000 residential customers receive electricity service directly from Santee Cooper.

POUs, primarily municipal utilities (or "municipals" or "munis"), were early advocates of policies to expand wholesale competition in electricity markets. Before the Energy Policy Act of 1992 and the Federal Energy Regulatory Commission (FERC) orders requiring open transmission access, most municipal utilities remained captive customers of the neighboring or surrounding investor-owned utility for their power needs. As long-term contracts with these local power suppliers started to expire, municipals became more interested in getting bids for power supplies from alternative providers and in getting greater access to transmission services that would make those purchases possible. After the passage of the Energy Policy Act of 1992, a number of municipals gained access to alternative suppliers, although some were subsequently saddled with having to compensate their former suppliers for stranded costs created by switching suppliers. (See Chapter 14 for a discussion of stranded costs.) Nonetheless, despite two major FERC rules calling for open access, a number of municipals remain frustrated in their attempts to obtain access to the transmission grid so as to obtain power from other suppliers. These municipal utilities look to the formation of truly independent regional transmission organizations as a necessary step to more open transmission access.

Rural Electric Cooperatives

The Rural Electrification Administration (REA) was created within the Department of Agriculture in 1935 to extend costly electricity service to rural areas by facilitating and financing the creation of customer-owned electricity cooperatives. In 1994, the REA was replaced by the Rural Utility Service, which has a broader focus on all utility services, including water and telecommunications, in addition to electricity. Today, there are more than 900 cooperatively owned utilities, most of which are distribution cooperatives that purchase electricity to sell to their customers. In some jurisdictions, such as the State of Georgia, groups of distribution cooperatives jointly own generation and transmission facilities to produce and deliver power for their members.

Federally Owned Utilities and Power-Marketing Authorities

Federally owned utilities and power-marketing authorities (PMAs) include the Tennessee Valley Authority (TVA), the electricity generating plants operated by the Army Corps of Engineers and the Bureau of Reclamation, and four federal PMAs. The TVA, the nation's largest producer of electricity, is a public corporation serving Tennessee and parts of six other states in the South. In addition to supplying electricity, the TVA provides flood and navigational control, agricultural and industrial development, and other services to the Tennessee Valley region. Originally, the TVA's focus was on hydroelectric power, and it invested primarily in dams that were built to generate electricity and to facilitate flood and navigation control. However, within two decades of its creation in the early 1930s, the TVA supplemented its hydroelectric generation with coal-fired plants and later constructed nuclear plants. By 1999, hydroelectric generation accounted for 8% of total TVA generation, whereas nuclear and fossil-fuel generation accounted for 30% and 62%, respectively.

The TVA services nearly 8 million people, mostly through 159 distributors (only municipals and cooperatives) in its exclusive wholesale service territory. These distributors receive power under long-term contracts requiring them to purchase almost all of their electricity exclusively from the TVA. In addition, the TVA sells power directly to 54 industrial customers and 8 federal installations.

The federal PMAs—Bonneville Power Administration, Southeastern Power Administration, Southwestern Power Administration, and Western Area Power Administration—market hydroelectric power produced by the federal dams operated by the Corps of Engineers and the Bureau of Reclamation. The PMAs sell power at the lowest possible rates primarily to *preference customers*, which are the POUs and cooperatives. Each PMA has its own geographic service area, statutory responsibilities, and transmission lines, except for the Southeastern Power Administration, which relies on the transmission facilities of other utilities to deliver electricity.

A Level Playing Field for Public Power?

Those who support competition in electricity generation and retail sales do so in part because they believe competition will make those markets perform better. Competition will encourage suppliers to minimize the costs of producing electricity, thereby lowering prices to consumers. Competition is also expected to yield the efficient amount of total electricity sales, that is, up to the level where the marginal cost of supplying one more kilowatt-hour of electricity just equals the market price. To achieve these goals, the rules governing the behavior of market participants should not place a cost advantage or cost disadvantage on any single competitor or group of competitors in the market.

The government applies different rules to publicly owned utilities and cooperatives than to investor-owned utilities. POUs and cooperatives receive various tax exemptions, financing preferences, preferential access to low-cost power, and regulatory exemptions not available to investor-owned utilities (IOUs). Collectively, these rules tend to impart a cost advantage to POUs and cooperatives that is not available to IOUs. We discuss these different advantages and associated challenges for restructuring in the following two sections: one on financial advantages and a second on regulatory exemptions.

Financial Advantages for POUs and Cooperatives

Tax Exemptions. Unlike IOUs, POUs are exempt from federal and state income taxes. Many of these tax exemptions also are available to electric cooperatives. In addition, POUs and electric cooperatives may be exempt from gross revenue taxes, franchise taxes, electricity sales taxes, property taxes, fuel taxes, and other taxes paid by IOUs. The financial effects of the exemptions from local taxes are muted somewhat in those regions where POUs are required to provide tax-equivalent payments as a substitute for tax payments at the local level. Beyond the required tax-equivalent payments, many municipal utilities contribute any revenues in excess of costs to the general municipal fund.

Tax-exempt status lowers the cost of doing business for POUs and cooperatives relative to the cost faced by IOUs. Ignoring other potential cost differences between the two types of utilities, it will tend to increase the share of electricity provided by POUs and cooperatives, because they can offer electricity at a lower price than IOUs with which they compete. Because the cost differential is only an artifact of the tax code, the increase in the POU's market share may displace lower cost IOU generation, raising the overall cost of supplying power.

There are also implicit benefits arising from the tax treatment of payments to cooperative members. Equity ownership in cooperative utilities entitles members, who are also the utility's customers, to a share of the revenues in excess of the cost of providing electricity service. Generally these payments, known as patronage dividends or capital credits, are increasing in the level of electricity consumption and are exempt from income taxes. These policies will lead to greater consumption of electricity by cooperative customers than among customers of similarly situated IOUs. With this practice in place, it would be difficult for IOUs or retail power marketers to compete for the business of customers of existing cooperatives.

Financing Preferences. POUs and cooperatives also have preferential investment financing options not available to investor-owned utilities. For publicly owned utilities, these options relate to the treatment of bonds used to fund the construction and maintenance of generation, distribution, and transmission facilities. The interest on publicly owned utilities' debt paid to bondholders is exempt from taxation. Tax-exempt bond financing creates a financial advantage for the bond issuer, because the interest rate on tax-exempt debt is lower than that required for debt with taxable interest payments. Explicit or implicit government insurance or guarantees on POU debt magnify the effect of these financial advantages.

Allowing POUs to use tax-exempt bonds to finance new investments in generating capacity could create an unfair cost advantage for this group of suppliers in a competitive-generation market. With this low-cost financing option, POUs could expand their investment in generation facilities and increase their generation market share at the expense of IOUs. As with the other tax exemptions described above, they would also sell more generation than similarly situated IOUs simply because of this artificial difference in capital costs.

Preferential Access to Federal Power. Power-marketing administrations are required by provisions of the Bonneville Project Act of 1937 and the Flood Control Act of 1944 to sell power generated at federal dams on a priority basis (mostly under long-term contracts) to a group of preference customers, which include POUs, cooperatives, military installations, and other federal agencies. The power is first sold to the preference customers at cost, with any excess available for private utilities. Certain federal laws specific to the individual PMAs may require power from a particular project to be sold to a specified buyer or within a defined geographic region. PMAs do not service the Upper Midwest or the Northeast, so access to federal power is not distributed equally across of the country.

Selling electricity at cost to a preferred class of customers creates inefficiency. If preferential access is maintained in a competitive world, the POUs and cooperatives could charge lower prices than IOUs and win more customers. In those markets served by POUs and cooperatives with access to federal hydroelectric power, the market price of electricity will tend to be low, leading to greater consumption of electricity than in other areas.

Regulatory Exemptions for POUs and Cooperatives

At the federal level, POUs and cooperatives are not subject to FERC jurisdiction and thus are technically exempt from the requirements of FERC Orders 888 and 2000. The "reciprocity provision" in Order 888 does require POUs and cooperatives to open their transmission lines for use by IOUs if the POUs and cooperatives transmit power over the IOUs' transmission networks. Also, though not required under Order 2000 to make the same filings regarding participating in regional transmission organizations as the IOUs, many public utilities, including some in the Midwest and Southeast, are teaming up with IOUs to explore forming large regional transmission organizations.

Although most publicly owned utilities do not own transmission lines, in the aggregate the POUs and cooperatives do own a large amount of transmission capacity. Public and cooperative ownership of transmission facilities complicates the creation of a seamless national grid to facilitate power trading among utilities and other market participants. As a result of the reduced opportunities to trade, wholesale power prices may be higher than they would be with a more open transmission grid. Having a patchwork of differently regulated transmission grid owners also limits the options for controlling transmission congestion (e.g., congestion pricing) and for dealing with system security.

POU participation in regional transmission organizations is complicated by the current restrictions on facilities use. Restrictions in the Internal Revenue Code limit the ability of POUs to use assets financed with tax-exempt bonds, such as transmission facilities, for private business use. If a publicly owned utility violates these "private-use restrictions," the Internal Revenue Service may be able to collect taxes from holders of the bonds or penalize the POUs issuing the bonds. The Department of the Treasury has issued some temporary rulings to exempt POU provision of transmission services from the private-use restrictions, but these rulings expire in 2004, so their implications for longer term commitments are highly uncertain. In addition, several clauses in state and local laws prevent POUs from selling, leasing, or transferring control of facilities to another entity, including an RTO.

POUs and Cooperatives after Restructuring

As the United States starts down the path toward greater competition in electricity generating and retail electric energy markets, several questions are being raised about the role of publicly owned utilities and rural cooperatives in a more competitive industry:

- Will these utilities continue to have access to tax-related advantages, including tax-exempt financing (and potential government guarantees) for their investments?
- Will transmission facilities owned by these utilities be subject to FERC oversight and rules?
- Will POUs and cooperatives be required to open their service territories to competitive suppliers?

- Will the private-use restrictions still apply to existing transmission and distribution facilities owned by POUs and cooperatives?

Many of these questions are being addressed in ongoing policy initiatives. At the federal level, legislation has been introduced in Congress that deals with both the continued use of tax-exempt financing for POUs and FERC oversight of transmission facilities owned by POUs and cooperatives. Under most state restructuring initiatives, POUs and cooperatives are allowed, but not ordered, to participate in retail competition unless they sell electricity outside their service territories. POUs and cooperatives are also preparing for a new electricity industry, and some have embraced competitive entry into their service territories.

Federal Legislation and Regulations

Both Congress and federal regulators have recognized the need to update and change certain rules governing publicly owned and cooperatively owned utilities to facilitate open transmission access and competitive markets. As was mentioned above, in January 2001, following a similar rule issued in 1998, the Treasury Department proposed temporary regulations regarding the private-use rules for electricity supply facilities financed with tax-exempt debt. In general, these rules broadened and clarified the definition of the activities for which these facilities could be used without violating the private-use restrictions. The activities included some sales of excess generation from existing facilities to private entities, allowing open access to existing transmission facilities and allowing POUs to surrender operational control of their transmission facilities to regional transmission organizations. These rules also clarified that, in general, new generation facilities built to sell into the market and new transmission facilities built for open access will be considered private use, and therefore will not be eligible for tax-exempt financing. These rules are still subject to modification, and they are set to expire in 2004. The 107th Congress is considering legislation that would make permanent changes of a similar nature to the rules governing what uses are and are not subject to the private-use restrictions.

The issue of whether transmission facilities owned by POUs or rural cooperatives should be subject to FERC regulation has also been addressed in different legislative proposals in earlier sessions of Congress. In most of these proposals, the extension of FERC authority over public power systems focuses on requiring open access from all transmission owners. They also propose a lighter form of transmission price regulation than is applied to investor-owned utilities, basically calling for FERC review of public power transmission rates and the authority to remand those rates back to the utility for revision when they do not meet with FERC approval.

State Legislation and Local Choice

The treatment of public power and cooperatives in restructuring laws and regulations varies across the different states. Public power utilities may elect to participate in retail competition under most state restructuring policies. These laws and regulations also usually include a reciprocity condition, which states that if a public power utility decides to sell to customers outside its service territory, it must allow tion. One proposal that has been put forth would require federal power to be sold at market rates to all customers. That plan would have competing private marketers submit bids to repackage and sell power generated by the Army Corps of Engineers and Bureau of Reclamation, eliminating the need for the PMA. The difference between the cost of the power and its market value would be split between the Treasury Department (to reduce debt) and the PMA generating region. This plan gives preference utilities the option to be the first to purchase federal power, but all transactions are at market rates.

Another proposal would give FERC the authority to alter the rates set by the Southeastern, Southwestern, and Western Area power-marketing administrations instead of just accepting or rejecting rate proposals, as is currently done. In doing so, the adjustments would be made to ensure that revenues are sufficient to recover the costs of providing electricity. In addition, transmission pricing for the PMAs would be brought under FERC jurisdiction, and all the PMAs would be subject to federal antitrust laws.

Who Should Decide?

The expansion of competition across the country suggests that the roles of publicly, cooperatively, and federally owned utilities will need to evolve to accommodate the changing market situation. Many of the policy decisions to effect this adjustment need to be made at the national level. Increased use of the transmission grid in a more open and competitive electricity market elevates the importance of a uniform and integrated approach to regulating transmission access and to overseeing transmission system reliability. This uniformity is best achieved if all of the transmission grid, irrespective of ownership, is subject to the same access rules and reliability standards. These rules and standards presumably would be set respectively by FERC and by the national reliability organization with oversight from FERC. The federal government will also need to rewrite the rules governing the tax treatment of public utilities and the role of federal PMAs and the TVA in the competitive marketplace.

Decisions about whether municipal utilities or rural cooperatives will continue to hold an exclusive franchise for retail electricity sales are perhaps best made at the local level. These decisions presumably will have direct effects on existing customers and few effects outside the local area. (For more information on the role of state and federal government, see Chapter 12.) Most state restructuring laws recognize this fact by allowing municipals or cooperatives within the state to decide for themselves whether to allow competing suppliers to sell electricity in their service territories. In states that allow retail competition, some municipals have opened their markets, but many have not. Some towns and city governments continue to believe that their municipal utilities will better meet the electricity demands of their citizens than will open, competitive markets.

competitors to sell inside its service territory. POUs that decide to allow competition are also allowed recovery of stranded costs, as are IOUs under most restructuring laws (see Chapter 14).

Restructuring legislation in many states also creates easier opportunities for cities and towns to get into the electricity supply business, through the aggregation of electricity demand within its jurisdiction. This option, sometimes called *community choice*, allows a city or town to purchase electricity on behalf of its citizens. The theory behind this is that, although individual residential customers may not have much bargaining power in dealing with power marketers, an entire town or city would be in a position to negotiate a good price. Under most municipal aggregation provisions, residents of a city or town engaged in aggregation are usually customers of the aggregator by default but have the choice to opt out and select a different electricity supplier. Conversely, a commercial retailer of electricity with many customers would have similar clout with energy wholesalers.

As restructuring creates uncertainties in electricity prices and reliability, some communities may decide to form their own municipal electric utilities. This option may be more attractive in states where municipal utilities are exempt from state regulation, other things being equal. For example, some municipal utilities interested in attracting industrial and commercial customers to promote local economic growth could give these customers lower prices than could be obtained in other locations. Whether it be for economic development purposes or to promote the local use of environmentally responsible power, communities will continue to examine this final option.

Federally Owned Utilities and PMAs

The federal government's role in the electricity industry is as a producer and marketer of wholesale electricity and as the owner of the TVA. Most of the power produced by the federal government, both inside and outside the TVA, is sold at the wholesale level to POUs or rural cooperatives for sale to ultimate customers. The TVA and PMAs own substantial amounts of transmission capacity as well.

The current status and treatment of federally owned utilities and PMAs raises three major challenges for policymakers seeking to define the rules that will govern their participation in a competitive electricity market. The first two issues, what to do about special financing arrangements and special regulatory treatment, are similar to issues facing municipals and rural cooperatives. The third issue, how to manage jointly the production of hydropower and other dam-related public services, is somewhat unique to federal hydropower facilities.

Financing

Similar to publicly owned utilities and rural electric cooperatives, the TVA and PMAs currently receive financial advantages over investor-owned utilities. The TVA does not pay federal, state, or local taxes, but the TVA Act authorizes and directs its governing board to make tax-equivalent payments to state and local governments. The TVA also has a high bond rating, owing at least in part to the implicit backing the federal government provides to TVA debt. This favorable bond rating, in combination with the tax exemptions on interest income for TVA bonds, reduces the TVA's cost of funds relative to those faced by an investor-owned utility.

The TVA's financial picture is not entirely rosy, however. Like many of its investor-owned counterparts, it has some costs that it might not be able to recover in a strictly competitive world, such as the costs of unfinished nuclear construction. This concern—coupled with an aging infrastructure and statutory limits on debt issuance that the TVA is close to reaching—are some of the weaknesses the TVA faces with the onset of competition.

The PMAs have been criticized for not recovering all of their costs in the prices that they charge for the power they sell at cost-based prices to nonprofit utilities. The Southwestern, Southeastern, and Western Area power-marketing administrations receive annual appropriations from the federal government for capital expenditures and operations and maintenance expenditures. The Bonneville Power Administration currently is financed by a revolving fund of revenue from power sales that is controlled by the Treasury Department. Critics suggest that not all of Bonneville's annual expenses are being covered by its sales revenues. Government lending to the PMAs at below-market interest rates and favorable repayment terms results in a net financing cost for the federal government. These factors create a potential competitive advantage for PMAs and their nonprofit utility customers over their investor-owned counterparts.

Regulation

Both the TVA and the PMAs set their own prices for both generation and transmission, and both are exempt from FERC regulation of their transmission rates and practices, including Orders 888 and 2000. The TVA governing board independently establishes both the wholesale power rates for electricity sold to distributors and the retail rates for power sold from the distributors to ultimate customers through long-term contracts. Federal laws have essentially created a "fence" that prevents power from being imported or exported outside of the TVA service territory.

Nearly a third of the bulk interstate transmission grids is not under the open access rules established by FERC. This portion of the grids includes more than 30,000 and 17,000 miles of transmission lines owned by the PMAs and the TVA, respectively. Excluding such a large amount of transmission facilities from FERC jurisdiction creates potential roadblocks to the long-distance transmission of power for wholesale trading and, thereby, can reduce the efficiency of competitive wholesale markets. Having a fractured grid can also raise reliability concerns, especially if the role of FERC as a final arbitrator of reliability concerns becomes more important in the future (for more information on reliability, see Chapter 11). The three transmission-owning PMAs voluntarily have filed open access transmission tariffs with FERC, which mitigate some of these roadblocks. In addition, the Department of Energy ordered the PMAs to participate in the development of regional transmission organizations and to make appropriate filings under FERC Order 2000.

Multipurpose Dams

The TVA's integrated system of dams and locks for managing the Tennessee Valley watershed was created to help promote economic development in the region and

to provide many public services, including flood control, navigation, resource stewardship, and the development of recreation areas, in addition to power generation. For the Western Area Power Administration, power-marketing activities are prohibited from interfering with the provision of irrigation services. These power-related activities must be addressed through an industry-restructuring policy.

The fact that federal dams provide multiple public services in addition to the "private" good of electricity complicates restructuring efforts. Because it is cheaper to have one facility provide all dam-related services, including electricity generation, than to have these services supplied separately, it seems unwise to preclude federally owned dams from selling electricity. However, it also seems unwise to allow tax-exempt federal dams with preferential financing arrangements to compete in a competitive wholesale electricity market.

Federal Policies for Federal Power

Current policies directed at federal utilities are not sustainable in a competitive environment. However, identifying a policy option that would be acceptable to all stakeholders is not easy. One option would be to eliminate the federal role in electricity generation—the competitive part of the business—through privatization.

Privatization would presumably remove all of the special tax and regulatory treatment currently enjoyed by federal utilities. It would also provide incentives for the utilities to behave efficiently. However, in the case of hydroelectric facilities, privatizing ownership and operation may not be consistent with promoting many of the other public services (e.g., navigation and flood control) that federally owned dams provide in addition to electricity generation. The commingling of private and public services from one facility make it difficult to separate the costs of each, but failure to do so opens the doors for charges of using public funds to cross-subsidize a competitive generation business and complicates achieving the full benefits of privatization.

As alternatives to privatization, several restructuring proposals in Congress would change how the TVA and PMAs operate. One proposal is to extend FERC jurisdiction over transmission facilities owned by the TVA and PMAs. This is a first step toward creating a uniformly regulated transmission grid that would facilitate efficient wholesale markets, the protection of reliability, and ongoing grid management.

In recent years, there also have been legislative proposals in Congress to essentially integrate the TVA into competitive markets by allowing exports and imports of power sales in its service territory. Under these proposals, the TVA would surrender authority to establish the retail rates set by its wholesale customers. Its long-term wholesale power contracts would also be revised to allow wholesale customers to purchase power from sources other than the TVA. In return, the TVA would be allowed to sell power in all competitive wholesale markets, including those outside its service territory, and to participate in some competitive retail markets as well. Proposals to open up the TVA service territory to competition generation include provisions for the TVA to recover any stranded costs. These proposals would also subject the TVA to some antitrust requirements.

There is also a debate about whether PMAs should sell hydroelectric power at the market price or at a price set as close as possible to its average cost of produc-

CHAPTER

14

Covering Stranded Costs

Before the California electricity crisis, perhaps the most highly charged (pun intended) issue associated with opening electricity markets to competition was whether and how to compensate utilities for the capital expenses they incurred during the regulatory era. If competition brings about lower prices, as its advocates would hope, utilities fear that they would not make enough money to recover some of these costs—hence, that they would be "stranded." The primary sources of such stranded costs, once estimated at upward of $135 billion, are associated with nuclear power plants and long-term contracts to purchase renewable and cogenerated power under the Public Utility Regulatory Policies Act (PURPA).

Utility advocates argue that a "regulatory compact" implicitly guaranteed cost recovery as part of the utilities' obligations to provide service. Those opposed to stranded cost recovery allege that utilities should not be rewarded for unwise investments and that forcing consumers to pay for stranded costs will thwart the objective of reducing electricity rates. In principle, deciding who is right should turn on a determination of whether regulators or utilities were in the best position to foresee restructuring and which of them were better able to adapt to the prospect of competition.

Stranded cost recovery has generally been part of the package necessary to build sufficient political support to implement the opening of retail markets. In addition, the federal government supports stranded cost recovery—perhaps not incidentally because the government itself is exposed by virtue of its ownership of electricity generation in the Tennessee Valley and Pacific Northwest. If stranded costs are recovered through surcharges on electricity purchases, it is important to devise methods that preserve competitive neutrality (i.e., not to introduce fees that create artificial cost advantages for either incumbent utilities or new merchant generators). Designing such a recovery system may be easier said than done.

Some of the issues associated with opening electricity markets are difficult because they are inherently complicated. Others, however, are difficult because the stakes are so high. Of these, the most intense issue surrounds the debate about recovery of *stranded costs*, that is, expenses that utilities believe they will not cover if, because of competition, they cannot sell as much electricity at the same prices as they could under regulation. Early estimates of these expenses ran up to and beyond $135 billion.

The battle over whether utility stockholders or electricity customers would be left to cover the stranded cost bill seemed sufficiently intense itself to thwart efforts to bring competition to power generation, even if the issues mentioned in other chapters here were trivial. The contentiousness of the stranded cost controversy is rivaled by the complexity of the issue when viewed from a disinterested perspective. Nevertheless, in most states that have seriously considered restructuring, stranded cost recovery has not stood in the way. When the State of New Hampshire tried to limit utilities to recovering only 50% of their stranded costs in its restructuring legislation, the utilities took the state to court and delayed the implementation of retail competition there by several years. Merits aside—and there are substantial arguments on both sides—recovery has proven to be politically expedient, and the magnitude of stranded costs has turned out to be considerably smaller than originally estimated.

What Are Stranded Costs?

At its core, the case for restructuring the electricity sector and to move from regulation to competition in generation is to reduce electricity prices overall. (Electricity prices may rise in some states, but those states' economies also would be better off overall; see Chapter 17.) Initially, one might have expected power prices to fall, for a variety of reasons. Lifting regulation should give the incumbent generators an incentive to reduce their costs through both improved plant management and adoption of more energy-efficient generation techniques. Open access makes it easier for independent power producers to build new, lower-cost plants using the latest generation technologies. Utilities in one part of the country where generation costs are low should find it easier to transmit, and hence to sell, power in parts of the country where prices have been high. Competition among these cost-cutting utilities, low-cost generators, and innovative entrants should translate into lower prices for consumers overall.

Presumably, without the promise of lower prices, restructuring would hardly be worth the trouble. A side effect of lower prices and competition, however, is that incumbent utilities would be getting less revenue from electricity sales from customers in the areas in which they formerly had exclusive monopoly franchises. In addition, competition probably means that some of their former customers will turn to independent power providers and low-cost utilities from other parts of the country. From a short-run perspective, the difference between these revenues and operating costs (e.g., fuel) represents a short-run profit from selling electricity. But under regulation, utilities would use such short-run profits to cover the long-run costs of constructing power plants.

This phenomenon is typical of most industries where suppliers have to incur substantial up-front costs in building factories, acquiring equipment, or establishing a brand name. In such industries, prices typically are just high enough to cover these costs over time, yet not so high as to encourage more firms to try to compete. But if demand falls unexpectedly, revenues can fail to cover the costs of investment made in anticipation of stable prices. For example, a firm might have invested $1 million in building a factory and business, only to find that there were not enough customers willing to pay high enough prices to make the money back. If, say, revenues provide enough of a surplus over operating expenses to recover $700,000 of those costs, we could call the remaining $300,000 "stranded."

Potentially stranded costs in the electricity sector come largely from three sources (on measuring these costs, see the box on the next page):

- Nuclear power plants that were either never completed or cost hundreds of millions or billions of dollars more than anticipated when initially planned. Furthermore, utilities are generally required by their regulators to pay the "decommissioning" costs of shutting down and safely dismantling the highly radioactive nuclear plants after their service life is over.
- Federal PURPA-mandated long-term contracts to purchase high-priced power at "avoided cost" from cogenerators or independent generators using renewable fuels (see Chapters 3 and 16). Some states determined that "avoided cost" of building new power plants was quite high, and thus the utilities purchased renewable or cogenerated power at high prices. To ensure that suppliers would earn enough over time to cover the cost of renewable power, the states required the utilities to purchase this power under long-term contract.
- Costs of other power plants that had not been fully written off.

In addition, utilities may require compensation if they are expected to continue to support a variety of public programs to encourage energy conservation, research into nonfossil fuels, and support of low-income electricity expenditures. The funding for these "stranded benefits" is discussed in Chapter 16.

Are Stranded Costs Shrinking?

As was noted above, early estimates of the level of stranded costs were quite noisy, but some reliable estimates (e.g., from Moody's Investor Services) ran upward of $130 billion in 1996. However, more recent estimates are considerably lower. The Department of Energy's stranded cost estimate in 1999—used to provide predictions regarding the effects of the Clinton administration's proposed Comprehensive Electricity Competition Act on electricity prices—was only $92 billion. Undoubtedly, the rise in prices in the western United States after the 2000 California electricity crisis reduced stranded costs even more.

The California experience illustrates the possibility that restructuring can create profit opportunities for generators as well as potential losses from stranded costs. Even before this situation, an extensive market in power plants arose due to the decisions of some utilities—mandated by some states, taken voluntarily in others—to divest generators. The results of these sales indicate that, despite their age, a util-

Measuring Stranded Costs: Lost Revenues or Unrecovered Investment?

Of the two methods for calculating stranded costs, the simpler is to calculate the revenues utilities would lose when they face competition. To calculate these lost revenues, one takes the following steps:

- Estimate expected prices under regulation and expected prices under competition, into the reasonably foreseeable future. Estimating these prices is not easy, as the price of electricity varies over an extremely wide range, depending on the level of demand, the ability of independent generators to meet that demand, and the capacity of transmission networks to import power from other regions.
- Estimate how much power the utility would sell, both under regulation and under competition. As the supply of power from competitors can vary with price and demand, this also is a tricky calculation.
- Multiply the prices in each year times the quantities in each year, to get revenues in each year, for both the competitive and the regulatory scenarios.
- Subtract the revenues under competition from the revenues under regulation to get the revenues lost because of competition.
- Calculate the present value of this lost revenue by discounting the revenue in future years by rates of return comparable to what investors could have earned in investments with similar risk characteristics.

This approach, which is based on estimating lost revenues, presupposes that the revenues the utility would have earned without restructuring would have been just enough, and no more, to compensate utility stockholders properly for the investments they made in generation. This may be false. If regulation continued, the revenues to the utility might be more than sufficient to compensate regulators for their investments.

A more accurate estimate of stranded costs would compare (in present-value terms) the amount utilities would receive after restructuring, over and above operating costs, to the level of investment in generation that had not been recovered (i.e., depreciated) before restructuring. That number, though more accurate, requires an estimate of the returns investors actually had received since building and expanding the generators currently in service. The "stranded cost estimator" thus not only needs to forecast prices and quantities—a difficult enough task—but has to know about how costs have been recovered in the past to determine how much exposure a utility's investors actually have.

ity's generators may be worth considerably more in a competitive marketplace than their value as carried on the utility's books.

An example can illustrate this effect. Suppose a utility put a generator into service 20 years ago at an initial cost of $300 million. Suppose that in the past 20 years, the regulator allowed the utility sufficient funds to write off $200 million of that investment, leaving an undepreciated "book value" of $100 million. Now, suppose that with retail competition, the market potential for that generation in both the area it was built to serve and new areas in which it can sell power leads it to have a market value of $150 million. This creates a value of $50 million for that plant, which is a profit to investors, and which could and should be counted as part of the overall recovery of stranded costs.

A final factor reducing the severity of the stranded cost problem is that much of the pain of coping with it has already been borne. Many of the states that were first to move did so because electricity prices were particularly high. Those are the states where generation expenses were likely to be among the highest, and therefore the states where stranded costs were likely to be the highest. As we will see below, in most of those states, settling the stranded cost question was a prerequisite for building a political consensus necessary to move retail competition from the drawing board to practical implementation. Some recent estimates from Moody's suggest that the level of stranded costs remaining to be resolved could be as low as $10 billion. Although still large, this number is a far cry from the $135 billion at stake just a few years ago, and well within the capacity of an industry with annual revenues in excess of $200 billion to absorb.

Why the Controversy?

In the economy at large, stranded costs are a normal feature of the ebb and flow of doing business. The possibility of losing money on investments is the downside of betting that one might make more than enough to cover one's costs. In a market economy, a government policy to provide systematic compensation for stranded costs would lead people to invest their time and money, even when the likely outcome is a loss. The investors would get to keep the profits from good decisions, but the taxpayers would pay the price for bad decisions. Other than situations when a social safety net is warranted because a severe downturn in a particular market "strands" communities and workers tied to a particular industry, the general policy that encourages the appropriate level of risk taking is to let the market determine profits and losses.

The presumption of letting the market chips fall where they may appears less compelling when the gains and losses are the result of government action. Examples abound:

- The government funds a scientific health-related research project, which leads to a discovery that renders a current drug obsolete.
- A state builds a new limited access highway that attracts most of the traffic and reduces the customer base for restaurants along the former main route.
- An agreement to reduce tariffs and promote international trade in shoes leads to new foreign competition that takes away business from domestic shoe suppliers.

In all of these cases, one might say that public policy decisions have caused costs to be stranded in the pharmaceutical, food service, and apparel industries. But, again, there is no presumption that the government should ensure that stranded costs are recovered.

The complication in the electricity context arises because the generation investments that could be stranded under competition were made while the utilities were regulated and overseen, primarily by state public utility commissions. Proponents of stranded cost recovery argue that those investments were made under a "regulatory compact." From their perspective, utilities built their plants to meet their obligations to provide electricity as defined and implemented by state regulators. In addition, their construction was undertaken in accord with regulatory requirements that all investments be "prudent."

The regulatory quid in exchange for the utility's quo, on this account, is to honor an implicit if not explicit commitment that the investors be allowed to earn a fair return on that investment. Opening markets to competition purportedly takes away such an opportunity. New entrants are not saddled with the costs of nuclear plants, long-term PURPA contracts, and older undepreciated plants, allowing them to offer electricity at prices below regulated rates. The government should honor the "regulatory compact" through other means to ensure that utility stockholders recover these stranded costs. Not only is failure to recover stranded costs unfair to utility investors, in this view. Failure to compensate utility investors for stranded costs could send a signal to potential investors that subsequent policy choices might prevent them from recovering those investments.

Opponents of stranded cost recovery are not without responses. In effect, charges to cover stranded costs will come from charges that are akin to general taxes on the generation, transmission, or distribution of electricity. Such taxes inherently reduce both the electricity customers' desire to use power and the willingness of new and potentially more efficient power producers to enter the market (see below). In addition, a guarantee of stranded cost recovery will reduce the incentive of regulated firms to avoid building uneconomical plants. With regard to fairness, opponents of stranded cost recovery claim that the utilities ultimately chose what investments to make, and thus should be responsible for them.

Collecting the Money

Deciding that utilities should recover stranded costs after power markets are opened to competition is only the beginning. The next step would be for legislatures and regulators to implement the policy by instituting a method to raise the money.

From the standpoint of economic efficiency, the ideal method for raising funds to pay off stranded costs would be through *lump-sum* charges to ratepayers. A lump-sum charge is a fixed amount independent of how much electricity one actually would be using. The advantage of a lump-sum charge is that it allows the price of electricity itself to remain based on the cost of generating it. An example of a lump-sum method would be to charge each household and enterprise a fixed fee based on an appropriate multiple of an earlier year's bill or power usage. Because the charge would be based on past conduct, it would have no influence on future power prices and would not result in any artificial, inefficient reduction in purchases.

Although lump-sum charges are conceptually pristine, they tend to be politically unappealing. Assessments of charges unrelated to use frequently lead to complaints regarding unfairness. Moreover, they would likely tend to politicize the process of stranded cost recovery. Everyone has an incentive to lobby to get their charges reduced at the expense of everyone else. Who wins and who loses that political fight is likely to be unrelated to ethical or social criteria for distributing the burden of cost recovery.

In theoretical terms, the next best method to recover stranded costs would be to raise the funds out of general tax revenues. In principle, the tax system offers the broadest range of options for raising funds so as to minimize the economic costs of the stifling of demand that taxes inevitably bring. But taxes suffer from political handicaps—first, because the word "tax" is anathema to many. Second, many feel that electricity-related costs should be covered through electricity-related fees, rather than through taxes paid regardless of one's electricity use.

Consequently, pragmatic considerations will lead to having stranded costs recovered from charges to electricity users. If so, two efficiency principles should be kept in mind in deciding at what times of the day and year stranded cost recovery fees should be imposed. The first of these is minimizing artificially reduced demand. As a rule, the more customers cut back power use in response to a fee—what economists call the *elasticity of demand* (see the box titled "Elasticity and Price–Cost Margins" in Chapter 9)—the lower should be the stranded cost markup. This principle, derived from a theory known as *Ramsey pricing* (see the box in

Chapter 8 titled "Basing Fees on Willingness to Buy: Ramsey Pricing") is designed to collect a given amount of money through surcharges so as to minimize the losses to the economy as a whole as those surcharges stifle purchases of power.

Competitive Neutrality

A second, more visible policy principle is *competitive neutrality*. A policy to cover stranded costs should not be implemented so as to create a false competitive advantage for one power supplier over another. In other words, customers should pay the same stranded cost markup regardless of the generator from which they purchase power. This policy avoids a situation where generator A would be the low-cost supplier without the stranded cost surcharge, but for some reason generator B would be the low-cost supplier after the surcharge.

The main concern regarding competitive neutrality has to do with the market advantages of the incumbent utility power provider relative to new entrants. The concerns can cut both ways. Suppose the stranded cost surcharge is structured so that one can avoid paying the cost simply by switching to a new generator or, for industrial users, by going off the grid altogether by generating one's own power. This response—what people in the industry call *bypass*—implies that people are leaving the incumbent only because of the fee, and not because alternatives actually come at a lower cost.

Accordingly, policymakers want to make the stranded cost fee *nonbypassable*. One example of a nonbypassable fee would be to fund the stranded cost payment through a tax on electricity used, regardless of whether it was provided by the incumbent, a new entrant, or self-generation. If all customers are still on the grid, a surcharge on fees paid for distribution is competitively neutral if it is paid whether or not one purchases electricity from incumbent generators.

A third option that could promote competitive neutrality is called a *shopping credit*, which is the difference between what customers pay the incumbent utility if they continue to purchase power from it and what they pay the utility if they purchase power from someone else. The reduction in payment, or "credit," is what consumers get when "shopping" for power from a competing generator. The remaining payment includes the fees for transmission, distribution, and contribution to stranded cost recovery. In effect, the credit becomes the net price of purchasing power from the utility.

The competitive neutrality principle should still hold—a shopping credit should be set so that whether a generator is dispatched depends on its cost characteristics, and not on whether it happens to be owned by the incumbent or an entrant. In practice, however, implementing a shopping credit may be difficult. If the shopping credit is set too small, customers must pay a relatively large price to the incumbent if they switch to new suppliers. Accordingly, customers will stay with the incumbent even if an entrant is a lower-cost supplier. Conversely, if the shopping credit is set too large, new firms may have an artificial market advantage. Any inclination to promote competitive neutrality as a goal may take a back seat to using the shopping credit to balance political pressure from incumbents, entrants, and customers.

A potential problem associated with implementing nonbypassable fees or shopping credits is that the size of the fee or credit itself depends on the intensity of competition. The lower the market price without fees or credits, the larger the stranded costs, and hence the larger the fee or the smaller the shopping credit. But this relationship need not reduce incentives to cut costs. If the electricity market is competitive, any single generator will take the price it gets—the market price less the nonbypassable fee, or the incumbent's price less the shopping credit—as given. A generator will get to keep any amount it saves in costs. The effects on entry are less neutral. A high nonbypassable fee or low shopping credit could have little effect if it causes electricity prices to rise overall, leaving entrants with the same profit margin on power sales. If the incumbent's electricity rates are fixed by regulation, these high fees or low credits will cut the prices the entrant actually sees, reducing its profits and making entry less attractive.

State policies may have two other implications for how stranded cost recovery can affect industry structure. One possibility is that under a state's franchising statute, a utility could no longer justify getting cost recovery on a generator if it took that generator out of service. A utility would then have an incentive to keep a generator in place, even if it were inefficient to do so, just to protect its claim on funds from stranded cost recovery.

A second possibility is that a state may be able to shift some of the cost of paying for stranded costs to taxpayers across the nation through a process called *securitization*, which turns the collections from stranded cost fees into "securities" by using them to back state bonds. The revenue from the sale of these bonds is used to compensate utilities up front for their stranded costs. Securitization reduces the amount the state has to collect because the interest payments on the bonds are exempt from federal taxation (see the box titled "Securitization" in Chapter 4). This reduces the size of the fees that a state needs to collect; the savings come, in effect, from the federal treasury.

Resolving the Question

The complications associated with designing efficient fees and ensuring competitive neutrality bring back to mind the question of whether stranded costs need to be recovered. If the legal obligations of a state and its regulator to cover such costs were clear, there would be no controversy. As a general rule, however, regulatory law does not require a guarantee of recovery, but rather a fair opportunity to earn a just and reasonable return on one's investment.

Largely for that reason, the regulatory compact cited as the justification for stranded cost recovery is what economists and lawyers call an *incomplete contract.* Not all terms and contingencies were or could be specified in advance. Among these was who would bear any stranded costs if the state or federal government decided to end generation monopolies and open power markets to competition.

When an unspecified contingency arises, it is the task of courts to figure out which party should bear the responsibility for dealing with it. A useful principle is to predict what the parties would have agreed to up front, had they found it worthwhile to do so. Such a principle tends to minimize rather than exaggerate the potential adverse consequences when provisions are left out of agreements.

In applying this principle, two issues are particularly important. The first is to determine which party would have been expected to have the best information about whether the unspecified contingency would arise. Generally speaking, assigning responsibility to the best-informed party makes sense, because that party will typically be in the best position to do something about the contingency were it to arise.

In the stranded cost context, the information question would be to determine whether the state or the utility was in a better position to predict that competition was coming down the road. If the utility knew, it would be in a better position to avoid building plants that would turn out to be uneconomical. If the regulator knew, it would be in a better position to order the utility not to build unwanted plants. A case can be made for concluding that either side might have been the better-informed party. Because retail competition decisions are ultimately the state's responsibility, the state public utility commission would seem an obvious choice. To the extent that competition is the result of national trends and technological developments, the utility may have been better placed to foresee what was coming.

The second question is which party would be better able to adapt to information about the contingency. In the stranded cost context, this question boils down to whether the utility was free to plan its generation capabilities in light of future competitive developments. If so, the case that it should be compensated for investments that proved uneconomical is rather weak. Such compensation would send a message to any regulated firm that carelessness in planning would have no adverse consequences.

Conversely, if the utility's decisions regarding plant construction and power purchases were made in response to regulatory pressure and against its better judgment, it may not have had the flexibility to respond to the possibility of future competition. In that case, responsibility for cost coverage should rest with the state. This may be more likely for some utility expenses than for others. For example, stranded cost recovery may be particularly appealing for expenses associated with long-term contracts to purchase electricity generated by renewable fuel, if utilities would not have signed those agreements but for state and federal mandates under PURPA (see Chapter 3).

Political Imperatives

When so much money is at stake, disinterested consideration of which party would have been more aware of forthcoming competition, or better able to adapt to it, may carry little weight. Experience with restructuring so far suggests that stranded cost controversies tend to be settled in favor of recovery, for reasons having as much to do with political expediency as they do with economic efficiency or fairness, as was noted above in discussing shopping credits.

First and perhaps foremost, restructuring is difficult, if not impossible, to bring off in a state without the support of the incumbent utilities. Opening markets requires a delicate balancing of the interests of a diverse set of stakeholders—the utilities themselves, their competitors, large industrial users, consumer groups, environmentalists, and state and federal regulators. The utilities are typically among the largest corporations and employers in a state, with a huge financial stake riding on the path that the electricity market takes. In addition, they own the distribution and transmission facilities that must be managed in an open, reliable manner for competition to be effective. Consequently, the success of retail competition will be more likely if the utilities see it as an affirmative opportunity. If they are financially threatened by competition, they may apply both legal pressure to halt moves toward opening markets and political clout to upset the potentially fragile coalition supporting such initiatives.

In addition, the government's interest in stranded cost recovery is not simply as an entity that sets policy in this area. The government itself is a participant in the industry. Estimates of the potentially stranded costs of federally owned generation in the Bonneville Power Administration and Tennessee Valley Authority run into the billions of dollars. The federal government may also face some stranded cost exposure as the guarantor of loans to rural utilities. Municipal utilities also may have some costs at risk if their markets open. Hence, the government has no small interest in promoting stranded cost recovery as a means to reduce pressure on their budgets.

One may view these developments with some cynicism, but it is crucial to remember that if restructuring was not a prospect—because, for example, technology would preclude effective competition among generators—ratepayers would continue to be responsible under standard regulatory procedure for covering these erstwhile stranded costs. Although regulators on their own might have attempted to disallow certain expenses associated with generation construction or power purchases, stranded costs would not have become a huge and widespread controversy without the likelihood that competition would do the "stranding." If the costs would have been paid without restructuring, perhaps policymakers should ensure that they will be paid with restructuring. Doing so would help competition to be seen as a way to improve electricity markets and the economy overall, rather than as an attempt to redistribute wealth from utility stockholders to consumers at large.

CHAPTER

15

Restructuring and Environmental Protection

Electricity generation is a major source of U.S. air pollution. In the midst of searching for new ways to reduce emissions, environmental regulators and other policymakers are eager to understand how increased competition in electricity markets is likely to affect the size of that sector's contribution to different air pollution problems. The effect of restructuring on the amount of air pollution will depend on three key factors: how competition affects the size of the market for electricity, how competition changes the mix of technologies used to generate electricity, and the form of environmental regulations governing electricity generators. In general, competition could lead to greater emissions of those pollutants that are not subject to strict caps, such as carbon dioxide, unless additional provisions are made to make it more attractive to use renewables and other technologies that emit few pollutants.

Opening electricity markets to competition is also likely to affect the performance of environmental regulation. Competition will likely limit voluntary actions to reduce emissions. At the same time, it will enhance incentives for electricity generators to take advantage of emissions trading. Environmental regulation and plant-siting requirements could reduce incentives for investment in new generation by potential competitors, which could adversely affect market performance. However, the magnitude of these effects is largely unknown and may be dwarfed by the investment-reducing effects of general uncertainty about the future of electricity restructuring.

The electricity-generation sector is a major contributor to air pollution problems in the United States. Emissions from electricity generators that burn fossil fuels contribute substantially to urban ozone and other pollution problems in U.S. cities, to acid rain in the Northeast, to regional haze and visibility problems in some rural areas, and to the buildup of greenhouse gases in the earth's atmosphere that has ignited concerns about global warming. Although virtually all electricity generators face some regulatory limits on their emissions of different pollutants, the environmental regulator's job is far from complete. Many U.S. cities are still not meeting their air quality goals for ozone and other local air pollutants identified as candidates for controls under the National Ambient Air Quality Standards in the Clean Air Act. In addition, the United States does not yet have a comprehensive plan to reduce greenhouse gas emissions to 1990 levels, the voluntary goal espoused in the U.N. Framework Convention on Climate Change that was signed and ratified by the United States in 1992.

In the midst of searching for new ways to reduce air pollution in general, environmental regulators and other policymakers are eager to understand how increased competition in electricity markets is likely to affect the size of that sector's contribution to different air pollution problems and to global warming. The answer to this question is not obvious. Restructuring will affect the nature of the electricity supply business in many different ways, some of which could lead to increases in emissions and others of which could lead to decreases. For example, on the one hand, competition is expected to lead to increased generation and emissions from older and currently underutilized coal-fired generators in the Midwest as they sell into new markets in the South and East. On the other hand, competition could accelerate the entry of state-of-the-art gas-fired merchant generators with low emissions that are seeking to make a profit in this newly opened market.

The transition to competition also could improve the performance of environmental regulations already facing the industry and of new regulations likely to be imposed in the future. Restructuring is focused largely on the generation arm of the electric power industry, which is the source of almost all of the pollution. Expanding competition will influence the effectiveness of environmental regulations and their cost. Competitive electricity generators face more pressure to reduce costs than do traditionally regulated generators, which generally are assured the recovery of all prudently incurred costs. As a result of this pressure, competitive generators are likely to respond differently to environmental regulatory requirements—in particular, to regulations that allow flexibility in achieving pollution reduction goals than would regulated generators.

One example of flexible regulation is an emissions *cap-and-trade* program, which places a limit, generally well below historic emissions levels, on the total amount of a pollutant that may be emitted from a set of regulated generators. Each generator is either granted or must purchase rights, also called allowances, to emit so many tons of pollution; each right typically covers 1 ton of emissions. Generators can trade these rights among themselves, and this trading generally lowers the cost of keeping total emissions below the cap. As we discuss in more detail at the end of this chapter, a generator in a competitive market reaps benefits if it can profitably cut emissions and sell permits and not bear the cost of having to buy more permits in order to emit more pollutants. Thus, it would be more likely to respond to the incentives created by a cap-and-trade program than it would under rate-of-return regulation.

Pollutants from Electricity Generation

Electricity generation produces several troublesome air pollutants. In all cases, the severity of a pollutant's effects on health and the environment depends on its concentration in the air, the length of time and intensity of exposure to it (e.g., for days rather than minutes, from inhalation while jogging rather than sleeping), and the sensitivity (e.g., of the lung or of a crop) to the dose. These pollutants include particulate matter (PM), sulfur dioxide (SO_2), nitrogen dioxide (NO_2), greenhouse gases, and mercury.

■ *Particulate matter*—soot, dust, dirt, and aerosols—has readily apparent effects on visibility and exposed surfaces. In addition, PM can create or intensify breathing and heart problems and has been linked to cancer and premature death. Power plants can contribute to PM concentrations in two ways: through direct emissions of particles and through so-called indirect emissions—emissions of nitrogen oxides and SO_2, which are transformed by chemical processes in the atmosphere into nitrate and sulfate particles of generally less than 2.5 microns in diameter (about 1/10,000 of an inch). Direct emissions of particles from power plants are effectively and inexpensively controlled with existing technologies, but indirect emissions remain a problem. Because small particles are believed to cause the most damage (although the most potent size and composition are still uncertain), the U.S. Environmental Protection Agency (EPA) has focused on what it calls PM_{10}, which are those particles in the air smaller than 10 microns in diameter. However, regulatory attention is focusing more and more on finer particles, down to 2.5 microns in diameter. Secondary particulates also contribute to regional haze and visibility problems in national parks.

■ *Sulfur dioxide* is a gas that may affect the heart and lungs in ways similar to particulates. In the air, some SO_2 (and nitrogen dioxide, described below) can convert into very fine particulates. In addition, SO_2 may damage trees and lead to acid rain, which can harm lakes and streams and also corrode exposed materials, such as the outsides of buildings. Title IV of the 1990 Amendments to the Clean Air Act placed a nationwide cap on annual emissions of SO_2 and permitted electric utilities (and other SO_2 emitters) to buy and sell the rights to emit SO_2 within the limits of the cap.

■ *Nitrogen dioxide* is a brownish gas with the potential to cause harm similar to that associated with SO_2. It also can convert to fine particulates, more so in the western United States than in the East; therefore, it is more relevant to regional haze problems in the West. Nitrogen dioxide is produced when nitric oxide (NO) emitted from power plants combines with oxygen already in the atmosphere. Many discussions of nitrogen-based air pollution refer to nitric oxide and nitrogen dioxide together as nitrogen oxides (NO_x), which can contribute to the acid rain problem in the northeastern states. In the presence of sunlight and volatile organic compounds, NO_x also can contribute to the formation of ground-level ozone (or smog), which causes respiratory problems and crop losses. EPA regulates ozone concentrations, but this standard is violated in more areas than are the standards for other pollutants. To reduce ozone formation, EPA has established a multistate summer cap on NO_x emissions in the eastern United States that is scheduled to take effect in 2004. NO_x emissions from power plants also contribute to nutrient pollution problems in surface water bodies such as the Great Lakes and the Chesapeake Bay.

■ *Greenhouse gases* consist primarily of carbon dioxide (CO_2), which is a by-product of the burning of fossil fuels. Greenhouse gases are thought to contribute to *global warming*, a general increase in the temperature at the earth's surface and the associated problems of climate change. The extent of both the global warming effect itself and the harm it may cause continue to be controversial. Nevertheless, there is a growing worldwide consensus that efforts should be made to reduce concentrations of greenhouse gases in the atmosphere. In 1997, the United States joined with more than 160 other nations of the world in negotiating the Kyoto Protocol, an international treaty that limits emissions of greenhouse gases from industrial countries. The Kyoto Protocol calls for the United States to reduce its annual emissions of greenhouse gases to 7% below 1990 levels by the 2008–2012 period. However, the U.S. Senate has not ratified this treaty, and early in 2001, President George W. Bush announced that the treaty is flawed and that his administration would not seek its ratification, but will instead develop an alternative proposal to deal with global warming.

- *Mercury* is emitted into the air from all power plants that are fueled by coal, oil, and municipal solid waste. Mercury exposure can lead to neurobehavioral dysfunction, such as tremors and blurred vision, and renal dysfunction. A developing fetus is particularly sensitive to the effects of mercury. Humans and other mammals are exposed to mercury primarily by eating fish from contaminated water. Emissions of mercury from power plants are currently unregulated; however, in December 2000, EPA announced its intention to regulate mercury emissions from power plants, and proposed regulations are expected by the end of 2003.

In addition to these major pollutants, electricity generation also can lead to the emission into the air of other toxic heavy-metal elements such as lead or cadmium.

In spite of far-reaching regulatory activity and major investments in energy efficiency and pollution abatement during the past three decades, emissions from power plants contribute substantially to the nation's NO_x concentrations, greenhouse gas emissions, and SO_2 emissions. Estimates of the electricity industry's contribution to the total amount of these pollutants in the United States are shown in Figure 15-1.

The Size of the Electricity Market

Allowing retail competition in electricity markets will have both direct and indirect effects on the size of the electricity market. In general, a larger market for power means greater emissions of all pollutants except those subject to an emissions cap.

A primary motivation for allowing competition in retail electricity markets is the expectation that competition, in general, will lead to lower prices for electricity consumers. If prices fall, demand for electricity and—holding the mix of generating technologies fixed—emissions from electricity generation can be expected to increase. Restructuring is also expected to induce greater reliance on time-of-day pricing, with potentially important long-run implications for patterns of electricity demand and emissions. The prospect for and realization of retail electric competition also is contributing to the demise of utility-sponsored conservation programs that have helped to slow the growth of demand for electricity in the recent past. Each of these three effects is discussed below.

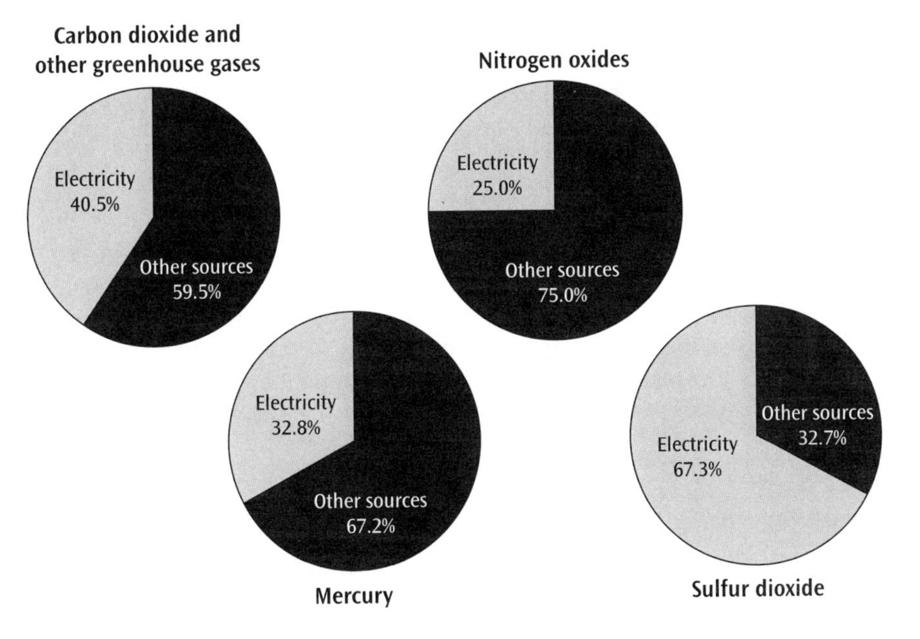

Figure 15-1

Share Contributed by the Electricity Industry to the Total U.S. Emissions of Four Pollutants, 1998

Sources: National Air Quality Trends 1998 (Washington, DC: U.S. Environmental Protection Agency). Mercury emissions data are for 1996; data are from *Mercury Report to Congress* (Washington, DC: U.S. Environmental Protection Agency).

Prices Overall

Whether and how much electricity prices fall as a result of restructuring will depend on a number of factors. If regulated utilities are relatively inefficient and the new market is very competitive and provides options for all classes of customers, then restructuring could produce substantially lower electricity prices. Conversely, if the local regulated utility is a low-cost supplier of electricity relative to its neighbors, then prices in the local area could actually rise under competition, particularly if new entrants are unable to beat the incumbent's price. This price increase could reduce local demand, with associated reductions in emissions. Recent evidence from California and elsewhere shows that the market price of electricity can increase after restructuring for other reasons, many of which are unrelated to restructuring. For example, increases in natural gas prices in 2000 and 2001 contributed to increases in wholesale electricity prices that—had they been passed through to consumers—could have reduced demand and emissions. (For more information on events in California, see Chapter 5.)

The size of a price decline associated with restructuring will depend in part on the level of stranded cost recovery allowed during the transition to a fully competitive retail market (see Chapter 14). The larger the amount of stranded costs that must be recovered through an electricity surcharge, the smaller the price reductions arising from competition in the short run. Over time, the impact of stranded costs on retail electricity prices will diminish as the contribution of stranded cost recovery to electricity prices becomes smaller and ultimately disappears.

In the near term, retail price reductions resulting from restructuring may be determined more by policy than by market forces. Many of the states that have already formally embraced retail competition have also enacted a guaranteed rate reduction or rate freeze that persists throughout a transition period of between 3 and 10 years. In these states, legislators, regulators, or both have guaranteed rate reductions, either by imposing a retail rate cap or by establishing a *standard offer rate* below the existing regulated price. This standard offer rate is the rate that will be charged to those customers who choose not to participate in the competitive retail market, and it effectively puts a ceiling on the price that a new entrant can charge to the eligible classes of customers.

Generally, these rate caps were initially set between 3% and 10% lower than regulated rates in effect just before restructuring, but some were set as much as 20% below regulated rates. Under standard assumptions about how responsive electricity demand is to changes in price, these price changes suggest that demand could rise by between 0.3% and 5%. However, the realized effects on usage may be much smaller; several states, including California, Massachusetts, and Rhode Island, have found that the retail rate caps or standard offer rates they initially set were too low to cover the costs of wholesale power purchases, and they have subsequently raised them.

Time-of-Day Pricing

Competition could lead to greater use of time-of-day or real-time pricing of electricity at the retail level. This method of setting prices yields higher prices during periods of peak electricity demand and lower prices during off-peak periods. As consumers see prices rise during peak periods, they may choose to shift some of their electricity-consuming activities to off-peak periods when prices are lower. The effect of this peak-shifting activity on emissions depends on the composition of the generating capital stock. In those regions where baseload capacity is mostly coal and peaking units are fired by natural gas, emissions would go up as a result of peak shifting. But if the increase in emissions occurs mostly at night, its effects on air quality, particularly on ozone concentrations, may be less damaging.

Demand-Side Management

Competition is already leading to the demise of many utility-sponsored programs for energy conservation, which is also known as *demand-side management*. These programs have been credited with helping to reduce demand for power, thereby delaying the construction of new power plants. If these programs are truly effective in reducing electricity demand, eliminating them will lead to more electricity being generated and higher levels of emissions of noncapped pollutants. Many of the states that have implemented competition have made provisions for the continued funding of energy conservation during a transition period of three to six years, using funds raised through a competitive transition charge added to the regulated electricity distribution rate. (Policy questions associated with these programs are discussed in Chapter 16.)

Competition and the Mix of Generation Technologies

For any given level of electricity demand and fixed set of environmental policies, the effect of restructuring on emissions of noncapped pollutants will depend on what happens to the mix of technologies used to generate electricity. Industry observers have identified several possible outcomes. One scenario foretells that restructuring will reduce the penetration of zero-emitting (at least of nonradioactive air pollutants) technologies, such as nuclear and many forms of renewables, and increase the generation at older coal-fired generators without NO_x controls, all leading to dirtier air. (For a discussion of generation technologies, see Chapter 2.) Another scenario envisions new entry of merchant generators employing highly efficient, low-pollutant-emitting, combined-cycle gas units and strong market demand for power from renewable generators, leading to a cleaner fleet of generators and lower emissions in a competitive world. Other mixed scenarios have also been suggested that would have more ambiguous net effects on emissions.

The effect of competition on the technological composition of the electricity industry can be decomposed into four main components: prospects for nuclear generation; prospects for renewables; new markets for old coal; and the rate at which new gas-fired generators displace older, existing coal-fired generators.

Prospects for Nuclear Generation

Nuclear power is an important source of generation, accounting for just under 20% of the electricity sold in the United States. Although there are substantial environmental problems associated with disposing of nuclear waste and concerns about the potentially devastating effects of a nuclear accident, nuclear power plants do not emit any SO_2, nitrogen oxides, mercury, or carbon dioxide, the pollutants that are associated with fossil-fired generation. Thus, from an air pollution perspective, nuclear power is clean. But without a change in public policy and in public attitudes, no new nuclear power plants are likely to be constructed in the United States in the near future, so that the percentage will diminish as electricity demand grows and as operating licenses of existing plants expire.

Whether or not the contribution of nuclear power plants to total electricity sales diminishes more quickly under competition depends on the net contribution of two opposing effects. On the one hand, competition may lead some existing nuclear capacity to be retired early. In a regulated environment, most nuclear power plants would be expected to remain online at least until the expiration of their current operating licenses. At market prices, some nuclear plants will be unable to cover the costs of fuel, operation, and maintenance, or to meet safety requirements. Estimates of the annual amount of nuclear generation potentially subject to early retirement range from 40 billion kilowatt-hours (kWh) a year to more than 110 billion kWh per year, or between 6.3% and 17.5% of current levels of nuclear generation.

On the other hand, competition will likely improve the efficiency of nuclear power plants. Improvements that were already made in the late 1990s in anticipation of competition took the form of fewer unplanned outages or shorter downtimes associated with refueling and therefore increased generation at existing plants during the course of the year. As a result of these improvements, electricity output at the surviving nuclear plants rose by an average of nearly 10% between 1993 and 1998. Future efficiency gains could also result from reductions in operating and maintenance costs as nuclear operators seek to reduce their production costs and increase their operating returns. Higher returns will help to keep nuclear plants online longer.

The bottom line is still highly uncertain. A 1999 analysis of electricity restructuring by the Department of Energy forecast that increased generation resulting from future productivity improvements at existing nuclear plants would more than offset the generation lost due to premature nuclear retirements. Indeed, although more than 5,700 megawatts (MW) of nuclear generators retired in advance of license expirations between 1992 and 1998, no additional early retirements occurred in 1999 or 2000. Moreover, several existing nuclear plants have been sold to companies choosing to specialize in nuclear generation, which presumably expect to operate them at a profit. Allowing competition in generation may actually prove to be the salvation of the existing nuclear power industry rather than its death knell.

Prospects for Renewables

Renewable generating technologies, or simply renewables, include all forms of generation that make use of a nondepletable energy source. This category of generators includes hydroelectric power, solar thermal and photovoltaics, biomass, and geothermal and wind power. Like nuclear power, renewables (except biomass) do not contribute to emissions of air pollutants or of carbon dioxide. In most policy debates over the future of renewables, discussion has been limited to nonhydroelectric renewables because of the political difficulty of siting new hydroelectric facilities in the United States.

Renewables currently account for roughly 12% of all generation, and nonhydroelectric renewables account for only 2% of all generation. Nonhydroelectric renewables have been slow to penetrate electricity markets because of their high cost. If as expected, increased competition in electricity markets leads to lower electricity prices, then—in the absence of cost-reducing technological developments—renewables will be less likely to penetrate the market. As the industry becomes more competitive, traditional regulatory programs to support renewables have been reduced in number and in size. (The future of these programs and potential substitute programs is discussed further in Chapter 16.)

Competition also brings greater possibilities for differentiating service offerings that could provide a boost to renewables. In states that have moved to competition, renewable generators and power marketers are developing service packages featuring *green power*—that is, power produced with renewable energy sources that, by definition, do not contribute to air pollution or nuclear waste. Under these packages, customers contract for power that is, for example, 20%, 50%, or 100% renewable, and generally pay a premium above the market price of conventional power. Some of these service packages are limited to nonhydroelectric renewables, but many are not.

To make it easier for customers to compare competing suppliers, including green producers and traditional power providers, on the basis of features other than price, several states are requiring electricity retailers and wholesalers to disclose information on the fuel mix used to generate their electricity and the emissions associated with that generation in a standard format established by a government regulation. The expectation is that this information will help give customers a better idea of what they are buying in emissions reductions when they choose to purchase green power (or, alternatively, what types of emissions reductions they are forgoing when they purchase from a traditional electricity supplier). The actual environmental improvements associated with this type of provision are likely to be small, because most electricity consumers will have an incentive to take a free ride on the green-power purchase decisions of others.

Whether green-power marketing ultimately increases, renewables generation depends on whether the size of the green-power market exceeds the contribution of existing renewable generators and on the selectiveness of green-power purchasers. Some green-power packages specifically indicate that a certain percentage of the power will come from new renewables, presuming that penetration above and beyond current levels is something that customers care about. Conversely, penetration of new renewables does not necessarily preclude early retirement of existing renewables that find it difficult to make money in a more competitive marketplace.

Ultimately, the contribution of renewables may be determined by policy instead of by the success or failure of private markets for green power. Restructuring laws in several states and several federal electricity bills include a renewable portfolio standard requiring that a minimum percentage of generation or power sales come from renewables, generally nonhydroelectric renewables. (This standard is discussed in Chapter 16.) The levels for the proposed portfolio standard range between 4% and 20% of total electricity sales or generation. If renewables did achieve a 20% share of all generation, sizable reductions in emissions of air pollutants from the electricity sector would result. Ultimately, however, more modest levels are likely to be adopted, and these will have smaller emissions-reducing effects.

New Markets for Old Coal

Many of the short-run economic benefits from more competitive generation markets will be realized through greater interregional electricity trade. Allowing generators to compete to serve distant customers may create opportunities for generation from low-cost older coal-fired facilities to displace generation from oil or natural gas facilities. This increased interregional electricity trade will result in higher emissions of CO_2 and other air pollutants. Competition will also give exporting generators incentives to improve plant availability, which could further increase the amount of electricity generated for export, and therefore the level of emissions. Conversely, generators will also have an incentive to economize on fuel use, perhaps by actually improving their heat rates, and this could contribute to some offsetting reductions in the emission rate per kilowatt-hour for carbon and other pollutants.

The extent to which interregional power trade will increase under competition depends significantly on the amount of available interregional transmission capability. Large differences in electricity prices between regions suggest at first glance that there will be greater incentives to expand transmission capacity under competition in order to exploit those price differences. In addition, the Federal Energy Regulatory Commission (FERC) open transmission access order nominally requires transmission-owning utilities to expand transmission capacity if necessary to satisfy a demand for service that cannot be met with existing capacity. But the incentives to expand transmission capacity essentially will depend on how transmission service is priced (see Chapter 8). If transmission is priced in a way that allows owners to earn excess profits whenever lines are congested, then they will have incentives to delay expanding capacity. Alternatively, if transmission users have rights to congestion revenues, the grid owners' incentive to delay investment in new capacity could be muted.

The potential for increased emissions as a result of greater electricity trading will diminish over time as older coal-fired generators are retired. Although, as a general rule, coal-fired plants have tended to outlive their original 30-year expected lives, these plants will not last forever. Indeed, some older plants will require capital investment to extend their lives, and the costs of these investments may not be recoverable in a competitive market unless the price of natural gas rises relative to that of coal. Increasing the output from older plants will also increase their maintenance costs, potentially making them uneconomic.

Older plants that undergo major modifications or substantially increase their production could be subject to the stricter emissions restrictions, known as new source performance standards, that apply to new generators. In late 1999, the Justice Department, on behalf of EPA, sued seven Midwestern and Southern utilities for making major modifications to 32 of their collective plants without installing the required pollution control equipment. During 2000, EPA settled with three utilities, two named in the original lawsuit and one additional utility that was vulnerable to future EPA enforcement action. In all the settlements, the utilities agreed to install additional pollution control equipment. If this enforcement action continues to be successful, it will raise the costs to all utilities of investments in life extension that may be necessary to continue to keep older, dirty plants online. New environmental regulations to limit emissions that contribute to the formation of fine particulates could also accelerate the retirement of older plants.

Market Penetration of New Gas

When England simultaneously privatized and introduced competition into its power sector, it also was phasing out price supports for the British coal industry. The result was a substantial penetration of new gas-fired generation owned and operated largely by new independent power producers. This switch from coal to gas resulted in both a reduction in carbon emissions and a dramatic reduction in emissions of SO_2.

Many analysts suspect that a similar phenomenon will occur in the United States. Driven by low natural gas prices and the advantages of gas-fired combined-cycle turbines (high efficiency, low cost, modularity, and the short lead time for bringing these units online), energy market entrepreneurs will see many opportunities to make money selling electricity with this technology and will seize those opportunities. In the late 1990s, dozens of energy companies announced plans to build new gas plants in California, the Midwest, and New England.

Despite these announcements, whether gas plants penetrate U.S. electricity markets faster than they would have in the absence of competition remains an open question, for four reasons. First, competitive markets are riskier for investors than regulated markets with more assured returns, and the cost of capital in a competitive market therefore will be higher than it would be under regulation. All else remaining equal, a higher cost of capital will tend to yield lower levels of investment in new generating plants.

Second, the siting of new power plants will continue to face regulatory hurdles, even in a deregulated environment. Concerns over the effect of power plants on environmental quality will have to be addressed before regulators will permit these plants to operate. In addition, new generating plants will need to locate in areas that have access to gas pipelines and to high-voltage transmission lines. The number of sites ideally situated for new gas plant development may be already occupied by existing generators, leaving a potentially limited number of sites for new entry by independent producers (at least in the short run). Third, uncertainty about the future path of natural gas prices could slow the penetration of new gas-fired generation. Fourth, the rate at which the industry can bring new gas-fired combined-cycle turbines on line may be limited by the capacity of existing equipment manufacturers to deliver the equipment. If demand exceeds their capacity to produce, this will bid up the cost of this equipment and could have a dampening effect on the rate of new entry.

The Form of Existing Regulation

Electricity generators face several different forms of regulation on air emissions that vary by pollutant and that can vary by location and by season when the emissions occur. The different forms of regulation include industrywide emissions caps on some pollutants, such as $SO_{2,}$, and maximum emissions rate standards for others. The form of restrictions on NO_x emissions from electricity generators can vary by location and by season.

For pollutants, such as SO_2, that are subject to an emissions cap, changes in generating activity resulting from the introduction of competition will not lead to any increases in total emissions, as long as the cap is binding. Under Title IV of the 1990 Clean Air Act Amendments, SO_2 emissions from electricity generators are regulated using a tradable-allowance approach. Each regulated generator must hold one emission allowance to cover each ton of SO_2 emitted from its facility, and total emissions of SO_2 from regulated facilities must be below the aggregate emissions cap. Between 1995 and 1999, the cap applied only to a subset of generators that pollute the most; beginning in 2000, the cap was tightened and extended to apply to all electricity generators.

Under the program, SO_2 emissions allowances are issued to generators based on a fraction of past emissions levels. Firms are free to sell these allowances if the market price exceeds their marginal cost of reducing emissions, or to bank allowances for use in future years. As a result of unused emissions allowances stockpiled between 1995 and 1999, a bank of more than 11.6 million allowances had accumulated at the end of 1999. During some or most of the next decade, emissions could exceed their annual capped level as the bank of emissions allowances is drawn down. Once the bank is exhausted, total annual emissions from this sector will not exceed the annual cap of approximately 9 million tons. Thus, expanding competition in the electricity sector will not lead to an increase in total emissions, except possibly during the early 2000s, when previously banked allowances are being drawn upon. Moreover, generators would be able to draw down this bank even without restructuring. Greater competition in electricity markets could lead to a change in the location of SO_2 emissions, however, which potentially could have adverse effects on local air quality in some regions.

Restrictions on NO_x emissions designed to address the acid rain problem take the form of maximum allowable emissions rates per kilowatt-hour generated. Under this form of regulation, changes in electricity production will affect emissions levels, but restrictions to address the long-range transport of ozone from the Midwest and South to the Northeast are being implemented as an overall emissions cap on generators in the eastern United States. Beginning in 2004, summertime emissions from electricity generators in 19 eastern states and the District of Columbia will be capped. This program virtually eliminates the potential for competition in electricity markets to increase total summer emissions in the eastern half of the country. However, emissions in other seasons and other regions could rise or fall, depending on whether competition increases or lowers production at facilities with high NO_x emissions.

Generator emissions of both NO_x and SO_2 will be further restricted if and when EPA implements its fine particulate standard. Because the air quality goals in this regulation would not become binding until after 2015, this regulatory development promises a longer term check on potential pollution increases associated with competition. The regulatory means for achieving required reductions in concentrations of fine particles, largely nitrates and sulfates, are very uncertain at this point, as is the precise timing of any future restrictions on emitters. In the meantime, several bills have been introduced in the 106th and 107th sessions of Congress to cap emissions of multiple pollutants from the electricity sector. Different bills include different groups of pollutants, but the most comprehensive proposal would cap emissions of NO_x, SO_2, CO_2, and mercury from electricity generators at levels substantially below current ones. These caps would prevent any increases in emissions that otherwise might be brought about by restructuring.

Emissions of CO_2 from electric utilities are currently not restricted. However, a large number of electric utilities participate in voluntary programs to reduce CO_2 emissions. In 1994, more than 500 U.S. utilities formally and voluntarily committed to reducing their emissions of CO_2 as a part of the Department of Energy's Climate Challenge Program. This program is credited with producing annual reductions in CO_2 emissions of at least 80 million metric tons for the past several years, with reported reductions topping 120 million metric tons in 1999. These commitments were made at a time when most utilities were subject to rate-of-return regulation and were therefore largely assured of recovering the costs of reducing emissions. In a more competitive environment, where electricity producers would have a stronger incentive to keep costs low, it is unlikely that such substantial voluntary commitments would be forthcoming without additional incentives or subsidies.

Putting the Pieces Together

Much remains unknown about how the different factors identified above will work together to determine the effect of competition on the electricity sector's share of U.S. air pollution. Existing research on this topic suggests at least three important lessons:

- First, even with a summer NO_x cap across the entire eastern United States, greater interregional power trading spurred by competition is likely to lead to some increases in annual NO_x emissions in other seasons in the East and in other regions. This effect will be strongest in the early years before older coal-fired units are retired.

- Second, the possible increases in national annual NO_x emissions likely to result from competition in the near term are substantially smaller than the reductions in NO_x emissions expected from the expanded eastern NO_x cap-and-trade program. The difference would be even greater if the NO_x cap were extended to 12 months instead of only the 5 summer months.
- Third, in the absence of new policies to promote the use of renewables, increased cogeneration, or greater investment in energy efficiency, competition is likely to lead to higher levels of CO_2 emissions from the electricity sector than otherwise would have occurred.

Relationships between Restructuring and Environmental Regulation

In addition to the magnitude and timing of emissions from electricity generators, greater competition also may affect how firms choose to comply with environmental laws. Without the presumption of cost recovery that existed under regulation, generators will be more concerned about minimizing the costs of environmental compliance in order to compete more effectively in the market. This new set of incentives may lead generators to make different environmental compliance decisions than they would have made under regulation. For one, in this more competitive environment, generators are also less likely to be willing to participate in voluntary initiatives to reduce emissions.

Yet competition will enhance incentives for electricity generators to take advantage of emissions trading. For marketable allowance programs to accomplish their purpose, generation companies holding emissions allowances must be interested in using them in the most profitable way—that is, generators must have a real incentive to lower their costs. Under traditional rate-of-return regulation, state public utility commissions often forced utilities to pass along to their customers, in the form of lower electricity prices, any profits from installing SO_2 abatement equipment and then selling excess SO_2 allowances. Under those circumstances, the utilities do not have much interest in selling allowances or reducing pollution beyond the reductions necessary to match their emissions with the allowances they are initially allocated.

In a deregulated market, however, where generators are free to earn as much as they can, the incentive to make these trades and to install low-cost abatement technology or switch to cleaner fuels is stronger. Therefore, we expect that in a competitive electricity market the advantages of emissions-allowance trading programs relative to command-and-control methods of regulation would be greater than in a regulated market. In light of these advantages, electricity generators may become stronger advocates for the cap-and-trade approach to environmental regulation, and this form of regulation may become more widespread.

Environmental regulation also can affect the extent of competition in—and thus the performance of—markets. In many regions of the country, such as the Southeast, generation markets are still fairly concentrated, with one or two generators dominating the market. In addition, in regions such as California, the Pacific Northwest, and the New York metropolitan area, available generating capacity is limited, creating the possibility of very high wholesale prices and even blackouts during periods of peak demand, particularly because retail prices generally do not reflect the true market price of generation at peak times. In both highly concentrated and capacity-constrained regions, entry by new generators is important to making generation markets more competitive. Entry will help to diminish the market power of existing generators in concentrated markets and help to increase the total amount of capacity available in the market during peak periods.

However, environmental regulations, including local siting restrictions and time-consuming approval processes, tend to raise the cost of entry and therefore may limit the number of potential competitors seeking to enter the market. (See the box, "Equity in Regulating Facility Construction," in Chapter 17.) To locate in some areas that are not in compliance with National Ambient Air Quality Standards, generators must purchase *pollution offsets*; these offsets—which can be very expensive—indicate that other formerly polluting sources have shut down or are reducing their pollution by an amount equal to the additional emissions expected from the new source. Siting requirements, which vary substantially from state to state, can increase the costs of building a new plant and lead to delays in bringing plants online. Uncertainty about the future course of environmental regulation—such as how and when the United States will decide to limit greenhouse gas emissions—can also have a similar investment-reducing effect that is independent of location within the country.

Whether the deterrents to investment created by environmental and siting regulations are substantial enough to have a big effect on the entry of new facilities is not known. It is conceivable that incumbent generators would try to use the siting approval process to try to block entry by potential competitors. However, uncertainty about the course and continued pace of deregulation may have an even bigger effect on current investment activity, particularly in regions such as California that had been slow to add new capacity in recent years. Also, when attempting to evaluate the potential competition suppressing effects of environmental regulations, it is important to keep in mind that the environmental and other benefits of these regulations could be substantial enough to outweigh these additional costs.

If electricity restructuring ultimately does lead to competitive markets, more efficient electricity production, lower electricity prices, and increased consumption, society will reap economic benefits. But increased consumption also is likely to have negative environmental effects, especially with respect to emissions of carbon. These higher emissions levels will make it more difficult to achieve specific emissions reduction goals. However, the expected efficiency gains in delivering electricity service and the greater effectiveness of incentive-based policies for reducing emissions could provide the means of affording a cleaner environment.

CHAPTER

16

Public Purpose Programs in a Competitive Market

Regulated electric utilities historically have undertaken several "public purpose programs" in addition to selling electricity. These activities range from offering rebates to consumers who purchase energy-efficient appliances—so-called demand-side management programs—to funding industrywide research and development of more efficient generating technologies. All have been made possible by the fact that regulators have, for the most part, allowed the regulated utilities to recover the costs of these activities in the prices that they charge electricity consumers.

In the newly competitive environment, utilities face greater pressures to reduce costs and, therefore, are reducing discretionary spending on optional activities, such as public purpose programs, that do not contribute directly to profits. At the same time, competition brings with it important changes in the incentives facing electricity suppliers and consumers that could eliminate or reduce the need for certain public purpose programs or require change in the means of provision to make them consistent with the reality of a competitive electricity market.

Public purpose activities traditionally funded by electric utilities on the policy agenda include demand-side management, use of renewable energy, research and development, and low-income support. Competition may affect the sustainability and implementation of these programs. However, competition also can change or eliminate the justification for public policies to promote these activities. New and proposed policies to promote public purpose programs may better suit a more open electricity market.

Regulated electric utilities historically have undertaken several *public purpose programs* in addition to selling electricity. These activities generally relate to the provision of electricity but have benefits that extend beyond the utility and the customers they serve. They include offering rebates to consumers who purchase energy-efficient appliances, that is, the so-called demand-side management

(DSM) programs; promoting use of renewable fuels; funding industrywide research and development (R&D) of more efficient generating technologies; and subsidizing electricity purchases by low-income households. Some of these activities have been imposed upon the utility by regulatory or legislative mandate, whereas others have been largely voluntary.

All of these activities have been made possible by the fact that regulators have, for the most part, allowed the regulated utilities to recover their costs in the prices that they charge electricity consumers. With the move toward more competitive markets for electricity, utilities' willingness to undertake or fund these public service activities is diminishing. Because utilities face greater pressure to reduce costs, they are reducing their discretionary spending on optional activities, such as public purpose programs, that do not contribute directly to profits. Either by changing the rules or changing their status (i.e., forming an unregulated generation company) utilities also are seeking to be free from such regulatory burdens as ratepayer-funded DSM programs or low-income subsidy programs that are not imposed on their nonutility competitors. As a result of these trends, and in the absence of new policies, the social benefits associated with these activities could be lost.

Yet competition also brings important changes in the incentives facing electricity suppliers and consumers. In some cases, these changes could eliminate or reduce the need for certain public purpose programs. In other cases, the justification for these programs remains unaffected by the introduction of competition, but the means of provision may need to change to be consistent with the reality of a competitive market.

This chapter describes four public purpose activities that traditionally have been funded by electric utilities: DSM, renewable energy, R&D, and low-income support. We discuss examples of each activity, and explain how each one is being affected by the advent of competition. We also review the justification for public policies to promote each activity and discuss how this justification could change with the introduction of competition. We conclude with a review of new and proposed policies to promote public purpose programs in a competitive environment.

Demand-Side Management

Demand-side management programs provide subsidies to electricity consumers for investments intended to reduce their energy consumption. These include:

- the provision of free home energy audits that provide information about how to reduce energy consumption in the home,
- subsidies toward the purchase of new highly energy-efficient appliances, such as air conditioners or hot-water heaters, and
- payments for allowing a household air conditioner to be retrofitted with a device that allows remote shut-off by the utility during periods of peak electricity demand.

State public utility commissions began actively promoting DSM programs in the late 1980s and early 1990s as a substitute for the construction of new generation facilities. A justification for these programs was that the regulated price of electricity to consumers, based on average cost, was often below the marginal cost of producing the electricity, particularly in peak periods or when the costs of pollution are not taken into account. The difference between marginal cost and price meant that, left to their own devices, consumers would have an insufficient incentive to conserve energy.

DSM programs can also reduce the economic effects of the incentive to overbuild generation capacity that can be created by traditional rate-of-return regulation. If a firm's allowed rate of return on invested capital exceeds its financial cost of capital, then it will have an incentive to overbuild generating capacity. Subsidizing DSM can at least partially offset the effects of this incentive to overbuild and will do so in an efficiency-enhancing way when the price of electricity is below its marginal cost or the cost of supplying the last kilowatt.

The suggestion that competition will preclude achieving the conservation and environmental benefits of DSM programs presumes that these programs really do reduce electricity consumption below what it otherwise might have been. In fact, the programs may reduce electricity consumption by less than program advocates and electricity regulators believe. To some extent, DSM programs subsidize conservation activities by electricity customers who would have made those energy-saving investments without the incentive payments, the so-called *free rider* effect. Second is the possibility that by lowering the cost of purchasing such energy services as heating or cooling, DSM programs could result in greater consumption of those services and thus a smaller fall in electricity consumption than would have occurred if consumption of the energy service had been held fixed.

Allowing competition in the electricity sector should help to solve some of the market failures that justify DSM programs. With competitive retail electricity markets, consumers will face separate prices for power and for transmission and distribution services. If electric energy markets are competitive, economic theory tells us that firms will price energy at its marginal cost.

Competition is expected to lead to more widespread use of time-of-day or real-time pricing of electricity, which should provide consumers with more accurate signals of the high cost of power during periods of high demand. Widespread use of time-varying prices will require substantial investment in new meters at the customers' premises. This will happen only if the energy cost savings associated with installing the meters exceed the cost of making these investments.

Another frequently cited but controversial justification for DSM programs is that consumers behave myopically when it comes to DSM investment. Numerous studies have shown that electricity consumers fail to invest in energy-efficient appliances that promise very high returns in the form of future savings. There are at least four possible explanations for this type of behavior. First is the suggestion that consumers do not have good information about the savings associated with a particular investment, and so they do not take advantage of this savings. Second, these information problems can be exacerbated when the person making the investment and the person reaping the savings are not the same. For example, landlords cannot charge higher rents for installing energy-efficient appliances that promise lower energy costs, and homeowners are not able to earn an adequate return on investments like installing superefficient windows when they sell their homes. Third, consumer failure to invest in energy-efficient appliances may reflect a rational choice in the face of uncertainty. Fuel prices could fall, making the expected returns from that energy-efficient investment risky. Fourth, the energy-efficient lightbulb perhaps is actually less preferable than a traditional lightbulb, and that is why consumers are not purchasing it, despite the savings.

Competition could help to eliminate the information problems consumers face regarding opportunities for energy-efficient investment. In competitive markets, new entrants and incumbent utilities should find it easier to market "energy services" by bundling electricity with appliances. Energy service companies could offer lighting, space heating and cooling, and other services instead of generic electricity. Marketing efforts by these companies should provide consumers with information on how much money they can save by buying bundled energy services and thus eliminate the need for them to calculate these benefits on their own.

Restructuring will not solve the problem of environmental externalities associated with electricity generation. However, as was discussed in the previous chapter, competition may improve the performance of emissions cap-and-trade regulations. If this mechanism is adopted to regulate a broader range of the pollutants associated with electricity production, the value of demand-side management programs as a second-best method of cutting emissions will be reduced.

Renewables

Utility programs to support the development and use of renewable generating technologies, which include biomass, geothermal, solar, wind and hydroelectric power, were largely a result of state and federal initiatives responding to the high prices of fossil fuels during the oil crises of the 1970s. Regulators and legislators wanted to encourage the development of alternative energy sources to reduce U.S. reliance on imported oil and on fossil fuels in general.

Toward this end, in 1978, Congress passed the Public Utilities Regulatory Policies Act (PURPA; see Chapter 3). This law required regulated utilities to buy all electricity generated by small renewable generators operating within their franchised service territory at a price equal to the *avoided cost*, the cost the utility would otherwise have had to bear to generate that electricity. It also allowed states to define how the avoided cost was determined. In some states, such as California, the avoided cost initially was set administratively at a time when fossil-fuel prices were near record high levels and it was assumed that they would not be falling. As a result, a lot of new renewable generating capacity came online to take advantage of these very favorable contracts. As time went on, California and other states used competitive bidding to define avoided costs, and renewables had difficulty competing with natural gas facilities on price alone.

Separate from PURPA purchase requirements, some state public utility commissions (PUCs) required utilities to obtain a particular amount of new renewables capacity or provided other financial incentives for renewables. For example, in the early 1990s the New York State PUC, as a part of its energy plan, ordered the regulated utilities to construct 300 MW of new renewable capacity. In 1992, the California PUC mandated that 50% of the new capacity constructed to meet growing elec-

ties may have little incentive to find new and better ways to reduce those emissions, even though such efforts could prove beneficial to society.

Whether incentives for research in this second category decrease or increase under competition depends on what happens with other policies and how electricity and related markets develop. For example, if environmental regulators expand the use of emissions cap-and-trade mechanisms to a broader set of pollutants (as was suggested in Chapter 15), electricity generators will have a self-interest in finding cleaner methods of generating electricity or less expensive methods of controlling pollution. When firms must effectively pay the price of an emissions allowance for each ton of pollution that they emit, they will have a direct incentive to try to change technologies to reduce those allowance costs. In combination with private markets for environmentally responsible power, should these prove successful, emissions cap-and-trade programs could also induce more R&D into the further development of cost-effective renewable technologies. Firms may still not be able to appropriate fully the results of these research efforts, but this problem is not unique to the electricity sector.

To the extent that electricity restructuring creates active markets for bundled energy services, competition could also create private incentives for conservation-related R&D. Energy service companies will profit by capturing some of the cost savings that their customers gain when they purchase bundled services instead of keeping their old appliances and purchasing their own electricity. In this environment, energy service companies or their appliance suppliers will be motivated to seek to improve the efficiency of the equipment they include in their retail service packages. Thus, private firms could have a stronger incentive to invest in R&D to improve efficiency than existed in a regulated setting.

In summary, competition is likely to lead to more R&D into process innovations and new products that promise higher returns to competitive firms. However, research that results in more public benefits will be less likely to be produced by the market, and public support may be in order.

Low-Income Support

Historically, utilities in a number of states have offered programs to promote "universal electricity service" to all households, regardless of ability to pay. These programs come in two varieties: programs to reduce the price of electricity and programs to reduce consumption of electricity. In 1996, utilities in 28 states offered programs to protect low-income electricity consumers, including affordable rates, discounted payment plans, percentage of income payment plans, and arrearage forgiveness plans. More states had policies governing utility credit and collection policies to protect customers who had fallen behind on their electricity bills. A number of states also historically have offered DSM subsidy programs specifically targeted at low-income consumers. All of these plans are paid for through revenues from retail sales of electricity.

The rationale for subsidizing electricity service for low-income customers has been expressed in two ways. First, providing an electricity subsidy to the poor promotes distributional equity across households of different income levels. Of course, a more efficient way to achieve the goal of greater equity would be to transfer money to the less well off, e.g., through tax credit and income assistance programs. A second argument for universal service programs is that everyone in the United States, whether rich or poor, has a right to available, reasonably priced electricity. This rights argument is also applied to other services, such as adequate nutrition, health care, and housing. Stated this way, electricity access itself, rather than alleviating the effects of poverty, becomes the direct policy objective.

If competition results in lower electricity prices for all consumers, then the affordability of electricity would be less of a problem for low-income consumers. Just how much less depends on how far electricity prices fall with competition. In some regions of the country that currently enjoy low prices, prices for consumers could rise if low-cost generators are free to sell into higher priced regional markets (see Chapter 17). The net effect of competition on prices for those who can least afford electricity could vary by location. In a competitive world, there will likely still be those who struggle to keep their lights on and air conditioning running. Thus, some type of program to subsidize electricity purchases by low-income households will continue to be desired.

Continuing Public Benefits Programs

The public debate over the issue of *stranded benefits*—that is, that programs addressing social goals would not be sustainable under competition—has two objectives: identifying which public benefit activities should be continued in a restructured industry and determining how they should be funded. The question of which of the four activities discussed above should continue to be funded through the sale of electricity has been answered differently in different states. Virtually every state that has passed a restructuring law or issued a final regulatory rule on restructuring has incorporated some subset of the four public benefit programs into that decision. Most states have chosen to fund these programs through the use of a nonbypassable "system benefits charge" assessed on electricity distribution. These charges range from 0.03 to 0.40 cents per kilowatt-hour (kWh). In some states, alternative or supplemental methods, including renewable portfolio standards, net metering, and generation disclosure, have been proposed to encourage the development of renewables.

Programs Being Funded?

Most of the states that have moved ahead with retail competition have required the establishment of energy efficiency and renewable energy programs. Many of the federal restructuring bills proposed in the 105th and 106th Congresses also included programs to promote energy efficiency and renewables in a post–restructured world. The size and structure of these programs varies across states and across different federal bills. For example, in Connecticut and Massachusetts, total annual funding for energy-efficiency programs alone is expected to be equal to about 3% of total revenues from electricity sales. In contrast, in Illinois, total annual funding for energy-efficiency programs is expected to be closer to 0.04%.

The least popular program among the states is public support for research and development. Only a handful of states explicitly fund R&D that is not focused specifically on renewable energy as a part of their restructuring efforts. The reluctance of states to fund R&D is not surprising, as the benefits would extend beyond the state line. However, in some states, a portion of the money earmarked for energy efficiency or renewable energy projects can be used (or even must be used) for R&D efforts in these particular areas. Maine has also established an R&D program supported by voluntary contributions from electricity consumers.

Nearly all of the states that had decided as of 2000 to implement retail competition have also chosen to have some type of low-income assistance program. These programs take a variety of forms, including bill assistance programs and programs to fund energy-efficiency investments for low-income households. Funding levels also vary substantially across the states, with some mandating that future funding be at least as high as current levels. A number of previously proposed federal bills also called for programs to make electricity more affordable for low-income customers. The most significant government-funded program for low-income households is the Low Income Household Energy Assistance Program (or LIHEAP), which is administered by the Department of Energy and which was funded at $1.4 billion in fiscal year 2001. This program provides block grants to the states to help eligible households meet their heating and cooling needs.

Funding Sources and Administering Programs

The almost universal mechanism that states are using to fund public benefit programs is the system benefits charge, also referred to in some states as the "public purpose charge." This is a nonbypassable charge levied on all retail electricity customers who take electricity off the distribution system. Associating the charge with the distribution of electricity, a regulated monopoly service, makes it virtually impossible to bypass. Even customers who generate their own electricity rely on the grid for backup electricity service.

The responsibility for administering these programs varies from state to state and across programs within each state. For example, in California the distribution utilities are responsible for administering the low-income program and the energy-efficiency program with oversight by the California Public Utility Commission (CPUC) and its advisory boards, the Low Income Governing Board and the California Board for Energy Efficiency (CBEE), respectively. Eventually, the CPUC and CBEE plan to select an independent body to run the energy-efficiency program. The renewable energy program and the R&D program are run by the California Energy Commission. In contrast, Delaware's restructuring law calls for the Delaware Economic Development Office to run the energy-efficiency program and the Department of Health and Human Services to administer the low-income assistance program.

The methods for dispersing funds also vary widely across programs and states. One approach worthy of note is the reverse auction that the California Energy Commission has used to select new renewable energy projects for funding. In this auction, qualified potential renewable projects submit bids on the size of a per-kWh subsidy, capped at 1.5 cents per kWh, they would like to be paid for all the renewable energy they generate. Projects are selected on the basis of their bid and other qualifying conditions. Winning bidders receive payments for all kilowatt-hours generated during the first five years of the project's life. Under this auction, payments are made for generation only, and thus much of the project risk is borne by the project developer and not assumed by the state.

Additional Programs to Support Renewables

The Renewable Portfolio Standard. Although most states use a system benefit charge to support low-income programs and energy-efficiency programs, several states have selected an alternative mechanism to promote the use of renewable resources, known as a renewable portfolio standard (RPS). The RPS typically requires that a minimum percentage of all electricity generated (or sold) within a region must come from renewable sources, typically excluding hydroelectric power. Generally, this percentage substantially exceeds the current contribution of renewable power. In some states and in all federal proposals that have included an RPS, the RPS is combined with a system of tradable renewable generation credits. Under this approach, each megawatt-hour of renewable power generated creates a renewable generation credit. A generator (or retailer, depending on which is responsible for meeting the requirement) can meet its obligation through some combination of generating directly, purchasing renewable generation under contract, and purchasing sufficient renewable generation credits and thus paying other electricity suppliers to use renewables to fulfill its obligation. This program is designed to allow the market to identify the least-cost way to satisfy the renewable obligation.

RPS systems vary dramatically along several dimensions. The first dimension is what types of renewables are included. Generally, the RPS focuses on nonhydroelectric renewables, but in Maine, for example, hydroelectric facilities and some types of cogeneration are included under the RPS. In some states, there are multiple standards, each applying to a separate category of renewables. The levels of the standards also differ across states and range from less than 1% a year initially in Nevada and Wisconsin to 30% a year in Maine. The Comprehensive Electricity Competition Act (CECA) of 1999 proposed a national RPS of 7.5% by 2010. Some states have minimum requirements focused specifically on new renewables, whereas others do not.

One of the appealing features of the RPS is the renewable credit trading market, described above, which allows for greater flexibility in meeting the minimum standards and therefore is lower cost. Credit trading will be allowed in some states (e.g., Texas). Other states (e.g., Maine) have decided to not allow credit trading until regional or national credit markets offering greater opportunities for savings and for sharing administrative cost develop. Even with credit trading, the cost of meeting the minimum RPS could be high, depending on the level of the standard. To limit the costs of this requirement, the CECA of 1999 included a cap on the price of renewable generation credits of 1.5 cents per kWh. The cap means that if the market price of credits exceeds this level, electricity retailers may simply pay the fee instead of purchasing additional credits in the market. In Texas, the price of renewable credits is effectively capped at 5 cents per kWh, the fine for failing to comply with the RPS.

Net Metering. Another policy to promote renewables is net metering. Net metering is the practice of allowing customers with small generating facilities that are interconnected with the local distribution company to install meters that run backward during periods when their generation is in excess of their demand. Net metering allows a customer to use her excess generation during one part of the billing period to offset her consumption during another part. If the customer generates more than she consumes in total during the billing period, then that excess generation is usually purchased by the distribution utility or the electricity retailer. In some states, net metering programs are limited to renewables, although other states also include small combined heat and power facilities and fuel cells in their net metering programs. This provision creates an incentive for electricity consumers to install small-scale on-site renewable generation, thereby reducing the need for generation from conventional sources.

Unlike the RPS, net metering is not a new idea born in the era of restructuring. Arizona and Minnesota have had net metering programs for small renewable generators (less than 40 kW in Minnesota and 100 kW in Arizona) since the early 1980s. However, several more states have introduced net metering requirements as they have moved toward competition, and net metering has been included in some federal restructuring proposals.

There is considerable variation across the states in terms of which renewables are eligible, which customers are eligible, what limits are placed on system size, whether or not there is a cap on overall enrollment, and what price is paid for excess generation. Most states limit eligibility to renewable systems under 50 kW in size, but open the program to all customer classes. In practice, states require only that net metering be offered by existing utilities and generally leave the decision of whether to offer net metering up to other electricity retailers. Before restructuring, most net metering provisions offered required utilities to purchase any excess generation from the net-metered customer at the end of the year. However, states that have restructured their markets, such as California, Maine, and Maryland, generally grant all of the excess generation at the end of the year to the utility or electricity provider.

Generation Disclosure. A final policy to protect consumers is the generation disclosure requirement found in several state laws and federal restructuring proposals. This provision requires that all electricity retailers disclose information on the mix of fuel sources and generation technologies used to generate the electricity they provide and the emissions of various pollutants associated with that generation. In most states, a standardized format for disclosing this information is being developed. Requiring disclosure of this information is intended to provide customers with the information necessary to compare the environmental performance of various electricity providers. Disclosure could prove difficult when power marketers purchase their power in a centralized wholesale market, instead of from particular generators. However, disclosure requirements should also help to reduce fraud in green energy marketing and thereby improve the credibility of all such green power marketers. This, in turn, could help promote the success of private markets for green power, providing a boost to renewable generation.

PART III

The Future

CHAPTER

17

Prospects for Restructuring

To help explain the issues involved in deciding if, when, and how to restructure the electricity industry, we have reviewed the technology, policy history, and several current state and international initiatives to institute competition in retail power markets. The California crisis of 2000–2001 brought forth a series of market, financial, and political calamities, with no shortage of explanations for the shortage of electricity.

The history of restructuring overall, as well as the California crisis, provokes a number of policy questions that we have tried to address:

- Why have competition in generation but not in transmission and distribution?
- Should utilities be involved in both the generation of power and the business of transmitting and distributing that power (and that of their competitors)?
- How might generators exercise market power, and what can the antitrust laws and competition policies do to control them?
- How should regulators set prices for transmission and distribution?
- What measures are necessary to keep the supply of and demand for electricity continuously equal?
- Can reliability be maintained as the industry moves from cooperating monopolies to contesting competitors?
- Should retail competition policies be left to states or set by the federal government?
- What roles should publicly owned power play as the electricity industry becomes more dominated by private enterprise?
- Will electricity competition hurt or help the environment and policies to protect it?
- Do policies to promote conservation and alternative fuels need to be adapted or amended as the United States moves from regulation to competition?

We hope that describing the different options helps citizens resolve these contentious problems as they consider whether and how to restructure electricity markets. Each issue is only a part of the larger "bottom line" question of whether restructuring is worth pursuing. After having reviewed these issues, we can offer some insights on the big picture itself by reviewing the advantages of markets overall and speculating on how some of the larger real or purported battles will play out. We end with what we believe is the largest decision our nation will have to make: Can the United States reap the benefits of deregulation without seriously risking the reliability of this crucial system?

Other Political Battles

Controversies over price spikes, retail electricity rates, rolling blackouts, and utility bankruptcies are not the only political battles facing policymakers and the public in deciding whether and when to restructure. Other major battles include industrial versus residential users, marketers versus customers, and high-cost versus low-cost states.

Industrial versus Residential Users

An oft-heard complaint is that restructuring will benefit primarily industrial users, with little or no benefits for households and small businesses. The basis for the story is that only the large buyers will have the expertise and clout to cut favorable deals. However, one should note that under regulation industrial users already get electricity at prices roughly a third below what households pay. Some of the discount industrial customers get may reflect real cost advantages in getting power delivered in bulk (e.g., having to spread billing costs over a larger volume of electricity) and being willing to tolerate scheduled power interruptions. Additionally, some industrial customers may be able to go "offline" and generate their own electricity if utility rates are not favorable.

The industrial discount may also reflect the political clout industrial users have been able to exercise in a system where governments rather than markets set prices. Through lobbying and perhaps threats to relocate, stop paying taxes, and put substantial numbers of a state's residents out of work, industrial customers may be able to get governors, legislators, and regulators to support lower electricity prices. A major advantage of opening retail markets to competition is that marketers can, in effect, aggregate the purchasing power of a large number of customers and replicate the kind of bargaining power that industrial users alone now have under regulation. Finally, even if industrial users do get discounts, competition among those users in their markets should ensure that those cost savings end up in the hands of consumers for those products.

Marketers versus Customers

The effectiveness of restructuring will also depend on how electricity customers feel about their ability to get what they want from those who sell them power. Some concern follows from the experience with telecommunications. A common reaction among the public is whether electricity competition means that their dinners will be interrupted by even more sales calls. Perhaps more concrete have been allegations of "slamming" (switching a customer from another provider to one's firm without the customer's consent) and "cramming" (billing for services that the customer did not select). Last and not least is a concern that trying to compare the price and service offerings of different providers is often daunting.

State and federal policymakers are well aware of these concerns. Some have proposed regulations that would require generators to provide consumers with a uniform presentation of price and pollution data, akin to the nutritional charts on food packages. Such data may help a consumer decide which generation company best matches his or her preferences for low price and environmental protection. State regulators, individually and collectively, are considering business practice rules regarding marketing and other business practices to prevent slamming and cramming. Regulators will also have to consider how to adapt cut-off protections—when a consumer who has not paid his bill loses access to electricity—to an era when generators, transmission companies, and local distributors all have separate claims to the customer's account.

A more daunting policy consideration is whether consumers would rather let the government choose their power provider and regulate its rates than to choose for themselves and rely on competition to set rates. Again, telecommunications offers some precedents. In a market economy, consumers presumably prefer having options, especially when competition among the possibilities leads to lower prices and better service. But in contrast to this presumption, a commonly heard complaint following the AT&T breakup in 1984 was that consumers were unhappy about having to pick a long-distance company, after years of having no choice.

Policies predicated on consumers' purported dislike of having to make choices need to be implemented with care. Reducing choices in the name of customer convenience can lead to more expensive and less desirable products sold to the consumers that such policies are intended to protect. We wonder how many consumers would prefer a return to the predivestiture days 20 years ago of just one telephone company, and whether they would currently like a world with only one cellular telephone supplier or Internet service provider.

Depriving consumers of options in electricity may preclude them from having their power company take a bigger role in managing energy use or from choosing sources that they view as being more environmentally benign. Still, consumer aversion to choice in electricity remains a phenomenon sufficiently significant to warrant the attention of regulators as they consider when and whether to open retail power markets. And, of course, consumer choice is meaningful only when they have more than one supplier from which to choose.

High-Cost versus Low-Cost States

Opening power markets, at either the wholesale or retail level, means that power companies in one state will find it easier to sell power in other states. The predictable outcome is that power prices will tend to become more equal across states, reducing enormous differentials in which customers in high-cost states (prederegulation California and New York) may pay double or triple the price for electricity paid in low-cost states (Kentucky and Washington). This price equalization can be a boon for electricity customers in high-cost states, but it may well lead to higher prices for power in low-cost states. People in the low-cost states will be competing with customers in the high-cost states for power; they will no longer be able to keep it for themselves.

In many ways, the issue is no different from settings where trade is opened up across different regions or nations. Opening trade generally makes a region better off as a whole. For there to be regional trade, importers need to get a product more valuable to them than the money they have to pay for it, while exporters need to get revenues exceeding the cost of producing the goods and services they sell. If either fails, the losing side will be unwilling to make the trade. In the case of electricity sold across state lines, the profits that power companies in low-cost regions would make from being able to sell at the higher national price will exceed the added payments consumers in those states would have to make to buy power at that national price.

In principle, the gains to the power companies could be redistributed to the customers in such a way as to make everyone better off. In practice, as with the profits created by regulations that make it hard to enter an electricity market (see box on next page), redistributing these gains in an effective and efficient manner will not be easy. Taxing in-state sales will only encourage exports, making matters worse for in-state consumers. Taxing exports can restore the status quo, but in doing so defeats the benefits to the nation as a whole from increasing access to low-cost power sources. Taxing profits themselves would likely discourage low-cost producers from expanding their operations and reducing the national electric power bill.

The Advantages of Markets

The growing intensity of the politics of electricity can obscure why we might want to go down that road. The promise of electricity restructuring is that a competitive market in power—accompanied by effective regulation of distribution and transmission and appropriate policies to ensure reliability—will lead to a more efficient electricity industry. Generators will have incentives to produce power at the lowest cost, and competition will allow those savings to be passed on to the consumers. Over the long run, if the many problems in maintaining load balances and reliability can be solved, opening markets should increase incentives to develop innovative power production technologies and service plans that will lead eventually to better service and lower prices.

The Profit Motive

Many find the profit motive that drives market behavior inimical to the public interest. However, the benefits that markets provide are driven by the quest for profits, in two related and important ways. The first is that business enterprises can make more money if they keep costs to a minimum through improved operations, capital investments, and innovative technologies that allow for greater savings down the road. Firms can also improve their bottom line through entrepreneurship, discovering and marketing new products and services that more closely meet the desires of consumers.

The second effect of the profit motive, perhaps paradoxically, is to ensure that these benefits eventually redound primarily to the public at large rather than to the firms themselves. The key is the process of competition. The profits that one firm makes through better operations, rapid innovation, and lower prices typically will attract others into the market as well. As these firms enter and expand their sales, competition among them should cause prices to fall. In principle, the efforts of firms to compete against each other to get profits from cutting costs and improving operations transfers these gains from producers to consumers.

The general success of markets relative to other, more centralized ways to organize and run economies has led them to be adopted around the world. Studies have generally shown that bringing competition to previously regulated U.S. industries has brought large net economic benefits, especially by creating an atmosphere in which entrepreneurial initiative and innovation can flourish. As we have discussed, ensuring that markets achieve their ideal goals requires attention to make sure that competition works well where it can and that price regulation promotes efficiency where monopoly is inevitable. In addition, policy can help improve the performance of markets by ensuring that the prices people pay reflect those social costs (e.g., pollution-related harm) and public benefits (e.g., basic scientific research) that might otherwise fall by the wayside.

The Role of Entry

Our experience with restructured electricity markets is still in its early stages. Some critics have emphasized the potential for market failure, in that restructuring has led or will lead to increased pollution, less support for conservation and renewable energy programs, and the exercise of market power by large power producers. As we discussed above, policymakers are or should be well aware of the need for policies to limit the scope of these market failures. Such policy goals include the need to prevent discrimination

Equity in Regulating Facility Construction

Making the trade-off between the health and amenity benefits of limiting generation and transmission construction and the potential cost in greater electricity rates should ultimately be the purview of the citizens affected by those decisions (see Chapter 12). Politically and economically, achieving a good balance between these benefits and costs is never going to be easy. Reliable and comparable measurements of the value of quieter neighborhoods and more pristine parks, and the value of cheaper electricity and fewer brownouts, are hard to come by.

Opening electricity markets to competition, however, may complicate this economic and political balance. Putting restrictions on the construction or expansion of facilities that increase the supply of electricity will raise its price, particularly in peak demand periods. As the text explains, all generators that sell power in the affected state or region will be able to charge that higher price. This increases both the amount of money consumers have to spend on electricity and the profits the generators get. The effect is what economists call one of *redistribution*. Although these restrictions may reflect the right balance citizens have between electricity supply and amenity benefits, they also "redistribute" wealth from consumers to generator owners.

This redistributive effect leads to an ostensible paradox—the electricity industry within a region supports restricting plant construction and expansion. At first glance, it may seem odd for an industry to support rules that seem to hurt it. However, if those rules restrict the ability of new firms to compete, they benefit the industry by helping to keep prices and profits higher than they otherwise would be. We may well find that if an industry is politically powerful, restrictions that it seems are designed to provide environmental benefits may end up as unduly stringent, because of the financial benefits they may bring to the industry itself. Citizens might be too reluctant to adopt otherwise good environmental regulations, because they do not want to see more of their income end up with power companies as a consequence.

In theory, a profits tax could alleviate these effects by providing a counterbalancing redistribution of wealth back to consumers. But the difficulty in properly attributing profits to the regulatory "windfall," as opposed to entrepreneurial skill and genuine cost advantages, implies that a tax could unduly discourage the efficient production of electricity. U.S. experience with similar taxes in the past—so-called windfall profits taxes on crude oil in the 1970s—supports this concern. Price controls on electricity could also redistribute wealth back to consumers, but we have seen that they may also lead to power supply crises in peak demand periods.

against independent power producers by transmission and distribution utilities (Chapter 7) and the need to limit the ability of any individual generation company to influence overall prices in any electricity market (Chapter 9). We also have looked at how restructuring is likely to affect air quality and the effectiveness of regulations to promote it (Chapter 15), and whether and how to revise policies to limit the use of electricity in general and of fossil fuels to generate it (Chapter 16).

An important factor in making this market (or any other) competitive is entry. Generators that earn excess profit should attract new generation capacity into the market. This new generation can help the problem in two ways. First, if prices during peak periods are so high that even marginal generators more than cover their costs, new peak-load generators would find it profitable to enter. As they come in, the supply of power at peak demand times would increase, driving electricity prices down until they just cover the cost of producing power.

Second, if baseload generators earn profits because they can now get peak period prices for their power at peak times, new generators would find it profitable to enter that market as well. This new baseload entry would drive prices down during off-peak periods. Prices could go as low as the bare cost of fuel and labor to keep these plants going during off-peak times, and prices during peak times could fall to the point that new generators would not find it economically attractive to come in.

The crucial role of new generation brings to light the importance of policies that can affect the availability of additional power sources. New power can come into a region in essentially one of two ways: through construction of new generation capacity within the region, including construction of new plants or additions to existing plants; or through expansion of transmission lines into the region. These would be particularly important during peak periods, when imports of power would be most valuable. As we saw in Chapters 6, 7, and 8, how transmission is structured and regulated as competition in generation is promoted is likely to influence the incentives to expand transmission capacity and make it available to any firm that wants to sell power in a particular area.

Regulation also can greatly affect the ability of power companies to expand the generators they already have in a region or to construct new plants. We are thinking here not so much of the traditional regulation of power as of the policies such as zoning and environmental protection that can make this construction more expensive and less expeditious—if not prevent it altogether. Policies that control the location and construction of power plants and transmission lines can clearly provide benefits. Few of us want a generator next to where we live or transmission lines running through our neighborhoods and parks. We only point out that the cost of these policies can be more expensive and less reliable electric power.

Lessons from California

The main question on the minds of observers who do not live in California is whether the situation in that state will be repeated elsewhere. Because the causes of the California crisis are many and varied—it is often compared with "the perfect storm"—predictions are difficult. The most crucial factor is the overall supply-and-demand situation in each state or region, taking into account generation capacity, fuel prices, and transmission availability on the supply side and population and economic growth on the demand side. These factors may have little to do with restructuring. The California experience, however, does offer some lessons, which those charged with implementing retail competition might consider.

Retail Price Controls

A first suggestion would be to lift retail price controls. If there is sufficient concern about retail market power to warrant retail price caps, because incumbent utilities or their spin-offs (see below) continue to dominate retailing and entry will not be forthcoming, continued regulation should allow pass-through of wholesale prices, particularly during periods of peak demand. If peak-period prices continue to be held below cost by regulation, policymakers may want to bring about some of the conservation that would have been induced by cost-based prices (e.g., via public demand-side management programs directed at peak-use equipment). If peak-period rolling blackouts cannot be restricted to relatively low-valued uses of electricity, policymakers may need to consider ways to encourage real-time pricing or its equivalent, including subsidized real-time meters and interruptible service contracts.

Market Design

A first step in improving market design would be to remove any regulatory impediments to long-term contracting between retailers and generators. To prevent the creation of incentives for anticompetitive conduct, long-term contracting should be offered first to retailers not affiliated with distribution utilities. If long-term contracting among unaffiliated retailers is not adequate, then perhaps distribution utilities could be ordered to divest their retail operations and become passive regulated wire companies. Such a divestiture could result in more than one retail company, limiting the need for retail price controls.

In considering the merits of direct dealing between the purchasers and suppliers of electricity, policymakers might want to discourage use of a signal power exchange. Instead, they might look at eliminating central auctions for all but ancillary services and information provision necessary to maintain system integrity. If such auctions need to be retained, they could consider requirements that a substantial minimum quantity be bid in at any given price or that a supplier must bid in all of its output at the same price to prevent gaming.

Real-Time Pricing

Regulators should consider eliminating any impediments they may have to allowing utilities to set power prices on a time-of-day basis. A major reason price spikes are so severe during peak periods is that, if consumers pay only average power prices, they will act as if power is just as cheap during those crucial times as it may be when the demand for power is not so large. Time-of-day pricing can encourage conservation, reducing wasteful uses of power during peak periods and the severity of price spikes. Active policies to encourage real-time metering are especially important if the alternative is to ration power through blackouts. Making users more sensitive to prices may also discourage generators from withholding output to raise prices.

Environmental Policy

It may also be necessary to factor higher electricity prices into the cost–benefit calculation in assessing environmental regulations involving pollution (along with the zoning and siting requirements discussed above). This should not be a license to ignore environmental costs, any more than the costs of building power plants, the fuel to feed them, and the labor to operate them should be ignored. But the benefits from environmental improvements should be balanced against the cost of making electricity more expensive to produce, and perhaps of making electricity markets less susceptible to competition from new suppliers. The environmental "lunch" was never "free," but the meal might be somewhat more expensive than was thought.

Market Power and Wholesale Price Caps

If the evidence supports the view that generators have market power, the Federal Energy Regulatory Commission or public utility commissions should be empowered to order additional divestitures to deconcentrate the market. The antitrust laws do not limit a firm's ability to raise prices by reducing output unilaterally. Regulation of output could be considered (e.g., through a "sick day" limit on outages suggested by Frank Wolak of Stanford University).

The dominant controversy is over the need to cap wholesale prices to limit market power. The best case for capping wholesale power prices is if generators can unilaterally exercise significant market power (Chapter 9) or if an auction conducive to placing high bids is in place (Chapter 5). Caps are not necessary to control collusion; agreements to fix electricity prices are illegal under the antitrust laws. Using a cap as a kind of windfall profits tax may be appealing, but the issue should be approached with caution. However well intentioned, wholesale price caps may discourage production and encourage consumption, putting the system at greater risk and turning a price spike into an emergency blackout unless generators are intentionally holding power off the market.

In addition, expectations that wholesale price caps are only "temporary" may not be borne out. Over the long run, generators would come into the market only up to the point where the revenues they expect would cover costs. As was noted in Chapter 9, this would imply that peak-period prices would be sufficiently high to cover capital costs as well as operating costs. For such prices to be high, supply will be relatively fixed. Theoretical and empirical analyses of the California markets indicate that a generator with even a small share of the market may have the incentive and ability to withhold output and raise prices (i.e., exercise market power), unless demand is more responsive to price than has been typically seen. If so, "temporary" price caps may be with us for longer than was originally intended.

The Key Problem: Reliability versus Competition

The most important cautionary note in emphasizing the severity of the California crisis is, ironically or perhaps paradoxically, that it may give a false sense of security regarding the merits of opening retail markets to competition. Most of the problems in California and elsewhere are ones we know how to solve or prevent. Antitrust laws, with additional deconcentration policies if necessary, can deal with market power. Mistakes in the design of residual regulation and centralized auctions can be avoided. But the most difficult items on the list—those dealing with supply-and-demand imbalances—will be with us, perhaps even more intensely, even if retail regulation continues.

In many cases, to be sure, implementing the best policy requires getting adequate data and balancing political interests, neither of which is easy. And often, markets can take some time to adjust, especially when the ultimate benefits depend on the construction of large capital projects, such as new generators and expanded transmission lines. But these are the kinds of problems we as a society have studied and addressed over many decades and in numerous contexts.

Thinking we have solved or can solve these problems could lead us to think that without these obvious (at least in hindsight) errors, electricity markets can work just fine. Perhaps that is so. Focusing on the rough, immediate issues that accompany any major industry transformation diverts attention from the deeper, longer term uncertainties that arise in bringing competition to electricity markets. These uncertainties come not from the similarities of electricity to earlier deregulated services, but from its differences. As we noted in the introduction, electricity is unique because of the confluence of three characteristics:

- *Critical to the economy.* Electricity is important not merely because the United States spends about 3% of its gross domestic product on power. A better measure of its importance may be to imagine how the rest of the economy and society at large depend on it. To take but one example, the defining adjective of the growth of the Internet as a business and communication tool is not "silicon" or "digital" or "software," but "electronic, as in "electronic commerce."
- *Technically exacting.* Many industries are crucial to the nation's welfare. A short list includes food, health care, housing, and transportation. But keeping supply and demand in constantly perfect balance is not a pressing concern in most of these industries. Excessive production may lead to some waste and costs in maintaining inventories. Too little production can lead to inconvenience—waiting for the next train, going to another grocer—and in some cases consumers can store a product to cushion the effects of any future shortages. But imbalances in most industries are more a nuisance rather than, as with electricity, a catastrophe, as we discussed in Chapters 10 and 11.
- *Interrelatedness.* If electricity were "only" critical and technically exacting, it would not necessarily present a potential crisis. If a supplier could not do a good job matching its supplies with its loads, it would be less able to attract customers in the future. The supplier could also offer warranties to insure their customers against losses resulting from power failures. Buyers could then choose among power suppliers on the basis of their reputations for reliability and warranties against outages. Because electricity suppliers and users are all on the same grid, however, the failure of one firm means failure for all. Consumers cannot protect themselves by picking the right supplier, and suppliers cannot ensure reliability merely by guaranteeing that they can cover demand from their own customers.

Electricity's critical importance, technical exactness, and interrelatedness distinguish it from other commodities. If the consequences of outages were not important, electricity policy could go on the far back burner once we adjust policies to the transformation from regulation to competition. If the electricity system were not fragile, policies to maintain load balances would not be necessary. If producers were not interrelated through the grid, producers and consumers could, through the market, arrive at the right combination of cost and reliability. Together, they will make reliability of the power system a continuing matter of public policy, even after we determine how best to deal with market power, air pollution, and other potential failures of electricity markets.

As a consequence, a high degree of coordination—through explicit cooperation, regulatory and legal incentives, or centralized management—is necessary to ensure that one supplier's imbalance does not bring down the entire system. A first question is whether competition is compatible with the centralized control necessary to prevent systemic breakdowns. Thanks in part to coordination among regulated monopoly utilities, U.S. power failures have almost always been local (e.g., when lightning hits a utility pole or local substation). Widespread failures, such as the New York City and Northeast blackout in 1965 or the western U.S. problems in 1996, are exceptional. Will this record continue as these utilities compete in each other's markets? If these newly competing firms cooperate to manage reliability, can fixed prices, reduced output, and divided markets be far behind?

Another way to promote reliability might be through the use of regulation and the law to hold individual generators responsible for systemwide losses incurred when they fail to meet the demands of their customers. Policy options could include regulatory reserve requirements and liability penalties when shortfalls in production relative to demand from one's customers lead to a blackout. Such rules may be ineffective if a generation company could declare bankruptcy rather than cover losses due to breakdowns. The time it typically takes the legal system to resolve liability disputes could be inadequate for electricity, where supply and demand must be kept equal without interruption.

If the need to ensure reliability requires central planning, the compatibility question becomes whether the role of that planner—a regional reliability council, regional transmission organization, independent system operator, distribution utility, or regulator of any or all of these entities—will leave sufficient scope for competition to be meaningful. If the central coordinator can limit its activity to relatively small and occasional purchases of ancillary services, the rest of the generation and marketing sectors will likely remain large enough to make competition worthwhile. Conversely, we have seen that operators of transmission systems and distribution grids become involved in imposing reliability requirements and mandating the provision of ancillary services necessary to keep the grid operating. In some cases, this leads them to become involved in the management of markets themselves, sometimes proscribing the independent contracting between buyers and sellers that drives most other businesses. The more the planner has to extend its reach into managing transactions, purchasing electricity, and owning generation, the more the scope of competition will shrink. In this respect, attempting to manage reliability in a market that is only partially regulated may be more difficult than when regulators had more extensive authority over the industry.

Whether the United States can reap significant benefits from competition in electricity markets, while retaining enough central coordination to maintain reliability, remains the most significant test for restructuring. Before the California crisis, the most likely threat to restructuring as a policy movement was a large-scale systemic breakdown followed by finger pointing, because no one would claim responsibility for ensuring reliability. Whether that will ensue, and whether it will impede restructuring, remains to be seen. The flaws in the California experiment imply that it is not a fair indicator of the prospects for restructuring. After all the institutions, regulations, and procedures necessary to ensure reliability are put in place, only experience will tell if there will be enough of a market left to have been worth opening to competition.

In debates about deregulation, advocates and opponents generally treat it as a matter of theory at best or ideology at worst. The electricity industry, more than most if not all others, asks whether the answer to "markets or not?" turns on facts as well as ideas and values. Market advocacy may be more effective if we concede that the issue is empirical rather than preordained. Electricity may turn out to join the list of other industries where deregulation has worked. But electricity could be the sector in which markets may have met their match.

Supplemental Reading

Befitting the importance and complexity of expanding competition in electricity markets, a vast number of books, articles, and Internet resources is available. Much of this literature is highly technical—particularly articles in academic economics and engineering journals on electricity system operation and institutional matters, such as auctions, transmission pricing, and market power. There are also numerous trade journals directed toward those in the industry, but these often come with high subscription fees and may not be available to many readers. However, some of the resources are accessible to a general audience. For those who want to investigate topics we cover in more depth or examine others, items on the list below may be helpful. This list is not comprehensive; it is only a starting point for additional inquiry—and we apologize for inadvertent omissions.

Books and Reports

Brennan, Timothy J., Karen Palmer, Raymond J. Kopp, Alan J. Krupnick, Vito Stagliano, and Dallas Burtraw. 1996. *A Shock to the System: Restructuring America's Electricity Industry.* Washington, DC: Resources for the Future.

A primer on electricity restructuring in the United States that introduces concepts and terminology.

Cooper, Mark N. 2001. *Electricity Deregulation and Consumers: Lessons from a Hot Spring and a Cool Summer.* Washington, DC: Consumer Federation of America.

A report, targeted to the general public from a prominent consumer group, that identifies problems of electricity restructuring (global and domestic) and presents policy suggestions to protect the interests of U.S. consumers.

Energy Modeling Forum. 2001. *Prices and Emissions in a Restructured Electricity Market.* Stanford, CA: Stanford University, Energy Modeling Forum.

This report—produced by a working group composed of members from government, universities, industry, and research organizations—examines electricity prices and emissions under a variety of competition scenarios to compare results from various models.

Fox-Penner, Peter. 1997. *Electric Utility Restructuring: A Guide to the Competitive Era.* Vienna, VA: Public Utility Reports.
Provides an overview of the complex policy issues, both economic and political, that surround electricity restructuring; much attention is devoted to the effects of restructuring on renewables and energy conservation.

Gilbert, Richard J., and Edward P. Kahn, eds. 1996. *International Comparisons of Electricity Regulation.* New York: Cambridge University Press.
Collects a number of political and economic case studies of electricity regulation and competition initiatives in Europe and South America.

Joskow, Paul L., and Richard Schmalensee. 1983. *Markets for Power.* Cambridge, MA: MIT Press.
This classic book, which is credited with bringing the possibility of competition in electricity markets to the attention of policymakers and academics, discusses the U.S. electricity industry structure and deregulation scenarios and related issues in the pre-restructured electricity industry era.

Stoft, Steven. 2001. *Power System Economics: Designing Markets for Electricity.* Piscataway, NJ: IEEE Press.
Covers price spikes, reliability, investment, electricity market architecture, market power, and transmission congestion issues related to the new generation markets and transmission organizations.

U.S. Department of Energy (DOE). 1999. *Supporting Analysis for the Comprehensive Electricity Competition Act.* Washington, DC: Office of Economics, Electricity, and Natural Gas Analysis, Office of Policy and International Affairs, DOE.
Describes and quantifies expected economic and environmental benefits of retail competition in conjunction with the proposed guidelines in the Clinton administration's Comprehensive Electricity Competition Act, submitted to Congress in the spring of 1999.

U.S. Federal Trade Commission (FTC). 2001. *Competition and Consumer Protection Perspectives on Electric Power Regulatory Reform: Focus on Retail Competition.* Washington, DC: U.S. FTC. http://www.ftc.gov/reports/elec/electricityreport.pdf (accessed October 29, 2001).
A staff report discussing the importance of competitive wholesale markets, demand responsiveness, reasonable residual retail rate regulation, and consumer protection in making retail electricity competition work. Extensive appendixes describe implementation of restructuring in 12 states.

Van Doren, Peter. 1998. *The Deregulation of the Electricity Industry: A Primer.* Washington, DC: Cato Institute. http://www.cato.org/pubs/pas/pa-320.pdf (accessed October 28, 2001).
Advocates a limited role for government in electricity markets, encouraging the use of competition rather than regulation to control transmission and distribution prices and letting businesses decide whether to unbundle wires from generation.

Periodicals

The Electricity Journal. Elsevier Science, New York; 10 issues a year.
Contains articles, news, summaries, and features suitable for U.S. electricity industry participants, federal and state government regulators, legal analysts, consultants, academics, and interest groups.

Public Utilities Fortnightly. Public Utilities Reports, Vienna, VA; 22 issues a year.
Geared toward industry professionals and government regulators. Provides articles, news, and editorials about all utility industries, including the electricity industry.

Articles and Research Papers

Borenstein, Severin J. 2001. Frequently Asked Questions about Implementing Real-Time Electricity Pricing in California for Summer 2001. University of California Energy Institute Working Paper. Berkeley, CA: University of California Institute. http://www.ucei.berkeley.edu/ucei/PDF/faq.pdf. (accessed October 29, 2001).
Defines real-time pricing and provides answers to such questions as, "If it provides advantages to other forms of retail electricity pricing, how much can be saved with real-time pricing, and who are the winners and losers under this pricing plan?" From the director of the University of California Energy Institute at Berkeley.

Burtraw, Dallas, Karen Palmer, and Martin Heintzelman. 2001. Electricity Restructuring: Consequences and Opportunities for the Environment. In *International Yearbook of Environmental and Resource Economics 2001/2002,* edited by H. Folmer and T. Tietenberg. Williston, VT: Edward Elgar. (Also, see Resources for the Future Discussion Paper 00-39, available at http://www.rff.org/disc_papers/PDF_files/0039.pdf).
Analyzes how electricity restructuring in the United States is likely to affect emissions from electricity generators and the performance of environmental regulation of the sector.

Goozner, Merrill. 2001. Free Market Shock. *The American Prospect* 12(15): 27–31.
Argues that electricity restructuring has failed to create truly competitive markets and that electricity producers have too much influence on price in deregulated electricity markets in California and throughout the country.

Green, Richard. 1999. Draining the Pool: The Reform of Electricity Trading in England and Wales. *Energy Policy* 27(9): 515–525.
Examines the conventional reasons behind the decision to replace the Electricity Pool of England and Wales and the proposed alternative structure of bilateral markets. The conclusion suggests that an efficient bilateral market could produce results similar to the pool.

Hogan, William W. 2001. Electricity Market Restructuring: Reform of Reforms. Paper prepared for 20th Annual Conference, Center for Research in Regulated Industries, Rutgers University. May 25, Newark, NJ. http://ksghome.harvard.edu/~.whogan.cbg.ksg/rut052501.pdf (accessed October 29, 2001).

Following a review of the roots of electricity restructuring, the author provides insights on competitive wholesale market structures and discusses newer reforms in England and Wales, New Zealand, and the United States.

Joskow, Paul L. 2000. Deregulation and Regulatory Reform in the U.S. Electric Power Sector. In *Deregulation in Network Industries: What's Next?* edited by Sam Peltzman and Clifford Winston. Washington, DC: Brookings Institution Press.

Provides a lengthy treatment on various aspects of deregulation and regulatory reform in the U.S. electricity industry. Includes details on the structure of state retail competition plans, divestitures, transmission network and wholesale market institutions, and the creation of wholesale power markets in California.

Joskow, Paul L., and Edward Kahn . 2001. A Quantitative Analysis of Pricing Behavior in California's Wholesale Electricity Market during Summer 2000. Cambridge, MA: MIT. http://econ-www.mit.edu/faculty/pjoskow/files/JK_PaperREVISED.pdf (accessed October 29, 2001).

Examines the increase in wholesale electricity prices during summer 2000 in California, finding a considerable gap between actual market prices and competitive benchmark prices even when controlling for increases in natural gas prices, demand increase, and decrease in power imports. Contains some technical discussions aimed at an audience with economics training.

———. 2001. Identifying the Exercise of Market Power: Refining the Estimates. Cambridge, MA: MIT. http://econ-www.mit.edu/faculty/pjoskow/files/exercise.pdf (accessed October 29, 2001).

Builds on the authors' previous paper cited above by revising data and responding to a research paper by other prominent analysts. Finds that California electricity prices exceeded a competitive benchmark in the summer of 2000, with strong evidence of the withholding of output from plants. Also contains some technical discussions aimed at an audience with economics training.

Rassenti, Stephen, Vernon Smith, and Bart Wilson. 2001. Turning Off the Lights. *Regulation* 24(3): 70–76.

Uses laboratory experiments to simulate electricity markets, finding that market power can cause prices to rise during off-peak periods with spikes on peak, but making demand responsive to price can counter these effects and reduce price volatility.

Speeches and Testimony

Duane, Timothy. 2001. California's Power Play Is a Crisis of Conventional Wisdom. Speech at the Energy and Resources Group of the Institute of Governmental Studies at University of California, Berkeley, February. (Available at http://www.igs.berkeley.edu:8880/research_programs/SACRAMENTO/duane.html; accessed September 26, 2001.)

An academic speaker provides additional insight into commonly cited factors contributing to the California electricity crisis while providing a list of principles that must be followed to develop a solution to the crisis.

Massey, William. 2001. Recent Lessons on Liberalization and Regulation in the U.S. Harvard Seminar on Energy Policy, July 6, Palma de Morca, Spain. (Available at http://www.ferc.gov/news/speeches/commissionersstaff/massey07-06-01Bspain.pdf; accessed September 26, 2001.)

Commissioner Massey of the Federal Energy Regulatory Commission discusses market design lessons, the importance of regional transmission organizations, and the need for aggressive regulatory intervention when markets fail.

Winter, Terry. 2001. Testimony of Terry Winter, President and Chief Executive Officer, California Independent System Operator Corporation Before the Subcommittee on Energy Policy, Natural Resources and Regulatory Affairs, House Committee on Government Reform, August 2, Washington, DC. (Available at http://www.caiso.com/docs/2001/08/02/2001080212092416892.pdf; accessed September 26, 2001.)

The CEO of the California Independent System Operator discusses the problems that have emerged in the California electricity crisis and what can be learned from past experience.

Websites

As with so much else, the Internet is becoming a preeminent source of information on electricity restructuring. Searching can turn up a vast array of material on government, industry, and academic websites. Among the most useful sites are the following sites.

American Public Power Association: http://www.appanet.org

Public power utilities are represented by this nonprofit service association, which advances the interests of its members and their customers. Although part of the website is available to members only, the association provides information on electricity restructuring, along with resolutions describing its positions on particular electricity industry issues.

California Independent System Operator: http://www.caiso.com

Information and data on the operations of the wholesale electricity industry in California can be found here. The available online documents range from memos to regulatory filings made with the Federal Energy Regulatory Commission.

California Public Utilities Commission: http://www.cpuc.ca.gov

This website for the state utilities regulator provides access to a variety of commission documents, including rulings, proposed decisions and resolutions, and general orders. The commission also includes a section devoted to California electricity restructuring with information on direct access and retail competition and electric service providers.

Center for the Advancement of Energy Markets: http://www.caem.org

The center is a nonprofit corporation with a mission including the understanding of changes in domestic and global energy markets along with developing market-oriented vision for future energy markets. Sample pages of the center's Retail Energy Deregulation Index, which measures a state's progress in adopting policies toward retail choice, are available online.

Edison Electric Institute: http://www.eei.org

This institute is the association of U.S. shareholder-owned electric companies and worldwide affiliates and associates. Although the information contained on its website is meant to be primarily accessible to its members, the institute does provide public access to a general information page covering a variety of topics and viewpoints in the U.S. electricity industry.

Electric Power Research Institute: http://www.epri.com

This institute is a nonprofit organization organized by industry participants to manage a comprehensive program of scientific research and technology development to serve its energy customers. Searches for descriptions of numerous technical reports can be performed on the website, with some available to nonmembers.

Federal Energy Regulatory Commission: http://www.ferc.gov

This commission—located within the Department of Energy—regulates the transmission and wholesale sales of electricity in interstate commerce and is responsible for licensing hydroelectric power projects and overseeing environmental matters in these projects. In addition to important issuances, speeches, and congressional testimony, many documents issued by the commission and many of the filings received from regulated parties are available online.

Harvard Electric Policy Group: http://ksgwww.harvard.edu/hepg/

This research forum—run by William Hogan of the John F. Kennedy School of Government at Harvard University—provides a forum for discussion and analysis of policy issues in electricity restructuring. Research paper titles are listed by subject group, and papers are available online or by special order. Summaries of seminar sessions are also available.

National Association of Regulated Utility Commissions: http://www.naruc.org

This nonprofit organization, whose members include state utility regulatory agencies, seeks to serve the public interest by improving public utility regulation. Included on the website is a map providing links to the websites of all state public utility commissions.

National Council for Science and the Environment: http://www.cnie.org/nle/eng-42

Summaries and links to additional information for subjects such as stranded costs and reliability are contained in this online electricity briefing book compiled by the Congressional Research Service with reports provided by the National Council for Science and the Environment, an organization supported by academic, business, environmental, and scientific organizations. Includes a section on the California electricity situation.

National Rural Electric Cooperative Association: http://www.nreca.org

This association is the national service organization dedicated to representing the interests of cooperative utilities and their customers through legislative representation before Congress, education, and training programs. Included on the website are news releases and facts about cooperative utilities.

North American Electric Reliability Council: http://www.nerc.com

The website for this council—a voluntary nonprofit organization of electric utilities devoted to developing, implementing, and enforcing electricity reliability standards—provides information on operating and planning standards, compliance, and reliability assessments. Many publications are available online.

Smithsonian Institution, Powering a Generation of Change: http://www.americanhistory.si.edu/csr/powering/content.htm

The Division of Information, Technology, and Society at the Smithsonian Institution's National Museum of American History, with funding from executives of leading power producers in North America, provides an introductory, chronological examination of the nation's electricity industry restructuring.

Steven Stoft: http://www.stoft.com

Stoft provides online chapters of his book, *Power Systems Economics: Designing Markets for Electricity*, numerous papers and talks on electricity, links to papers related to California electricity crisis, and links to a variety of electricity restructuring websites.

University of California Energy Institute: http://www-path.eecs.berkeley.edu/ucei/

This institute facilitates research and educates students and policymakers on energy issues important to California and the rest of the world. Its website provides online research papers from the Program on Workable Energy Regulation, which focuses on economic and policy questions in energy markets. Also included is a collection of data related to the California electricity markets.

U.S. Department of Energy, Electricity Restructuring Program: http://www.eren.doe.gov/electricity_restructuring/

This website provides a description of the department's Electricity Restructuring Program and an overview of electricity restructuring. Included are weekly updates on federal and state restructuring activities, along with an archive of previous updates back to 1996.

U.S. Department of Energy, Energy Information Administration: http://www.eia.doe.gov/cneaf/electricity/page/restructure.html

The department's statistical agency provides special studies, publications, forecasts, and data on the U.S. electricity industry. Many of the charts contained in the present volume are based on data provided by the administration.

Index

Adequacy of electric power, 6, 116
 of distribution, 118
 of generation, 118, 123–24
 See also Reliability of electric power
Air pollution, 18, 131, 159–60
 emissions reductions, 177
 pollutant types, 160–62, 163*f*
 regulation types, 169–72
Airline industry deregulation, 129
Amenity benefits, 129–30, 191
Ancillary services, 5, 19, 106, 108, 115
 emergency power replacement, 110, 111
 liability issues, 111
 load balancing and, 106, 108–13
 load following, 19, 109–10
 regulation of generation, 19, 109
 reserves, 19, 106, 111–12, 121
 responsibility for, 110–13
Antitrust laws, 5, 27, 63, 92, 105, 134
 market power and, 93–94
 See also Monopolies in electricity industry
Asymmetric information in markets, 63
AT&T breakup, 72, 75–76, 80
 See also Telecommunications industry regulation
Auction design and California crisis, 55–56, 57
Avoided costs, 28, 151, 176

Baseload demand, 17, 18
Bilateral contracting, 114, 115
Biomass power sources, 16, 18, 176
 See also Renewable power (nonhydroelectric)
Blackouts. *See* Outages and blackouts
Bonneville Power Administration, 23, 141, 142, 146
Bulk power reliability, 121–23
Business standards and practices, 134–35
Buyers of electricity. *See* Customers of electricity
Bypassable/nonbypassable charges, 40, 155–56, 181

California power crisis (2000), 3, 5, 9, 33, 46–48
 auction design, 55–56, 57
 background, 38–40
 causes (possible), 49–50
 consumers in, 52–53, 57
 contract sales in, 54–55, 57
 lessons from, 45, 192–94, 197
 market design and mechanics, 53–56, 134
 market power, 56–59
 metering (real-time), 53–54, 57
 Pennsylvania compared, 42
 price regulation and controls, 51–52, 57
 prices, 47*t*, 51–52, 97
 regulation and redistribution in, 51–53
 supply and demand issues, 50–51, 96
Cap-and-trade regulatory programs, 160
CCGTs. *See* Combined-cycle gas turbines
Chile and restructuring, 34–35
Climate change, 159–60
Coal-fired plants, 14, 17*f*, 18
 environmental issues, 160
 new markets, 167–68
Cogeneration, 8, 23, 28, 69
 stranded costs of, 151
Combined-cycle gas turbines (CCGTs), 15–16, 17*f*, 18
 market penetration, 168–69
Commercial customers of electricity, 22
 in California, 41
 in Pennsylvania, 43–44
 prices (U.S., 1999), 30
Competition in electricity industry, 1–10, 26
 antitrust laws and, 92, 93–94, 105
 competitive effects, 101
 deregulation and, 3–4
 drivers of, 31–32
 encouraging, 92–105
 environmental protection and, 8–9, 159–60, 178–79
 governmental roles, 6–7
 incentive regulation and, 85–86
 issues, 3–9
 load balance and generation responsibilities, 5–6
 mergers and, 92, 96–97, 101, 102–5
 monopolies and, 4, 94–95,
 public power and, 8–9, 136–38, 143–45
 public service programs and, 9, 173–74, 180–83
 regulation and, 3–5, 26–32, 61–62, 68–69
 reliability and, 6, 9–10, 187, 194–97
 restructuring and, 3–9
 stranded costs in, 151–52
 trends, 21
Congestion of transmission, 20, 87, 89–90, 91, 100, 143
 fees, 87, 90, 91

Conservation of electricity, 135, 178–79
Consumers. *See* Customers of electricity
Contract sales and contracting,
  bilateral vs. pool, 113–15
  in California crisis, 54–55, 57
  incomplete contracts, 156–57
Convergence mergers, 98, 103–4
Cooperatively owned utilities, 7, 22, 23*f*, 27, 136–38
  community choice, 145
  federal controls, 141–43
  financial advantages, 141–43
  industry functions by, 139*f*
  preferential power access, 142–43
  prospects, 143–45
  regulatory exemptions, 143
  rural electric cooperatives, 22–23, 140
  states and, 144–45
Cost-based regulatory systems, 30
Cost-of-service regulation, 82, 86, 98
Cost recovery. *See* Public service programs; Stranded costs
Costs (generation), 17
  avoided costs, 28, 151
  fees, 87–89
  fixed, 17, 85, 87, 93
  least-cost, 17
  marginal, 85, 87
  price–cost margins, 96, 97
  ramping, 17–18
  stranded, 8, 42*t*, 44, 149–58
  variable, 17,
  See also Stranded costs
Cross-subsidization, 75–76, 80, 95, 121
Customers of electricity, 6, 22
  adequacy of generation and, 123–24
  bilateral contracting, 114, 115
  in California power crisis, 41
  distributed generation and, 69–70
  equity and tradeoffs, 129–30, 191
  marketers vs., 188–89
  preferences, 188–89
  prices (U.S) by classes of, 30*f*
  stranded costs and, 150, 154
  subsidies to, 174–76, 179–80
  See also Commercial customers of electricity; Industrial customers of electricity; Residential customers of electricity

Decommissioning of nuclear plants, 14–15
Demand for electricity, 16–19, 83
  See also Customers of electricity; Dispatching generators; Generation of electricity
Demand-side management (DSM) programs, 9, 164, 174–76, 179
  See also Public service programs
Depreciation and rate regulation, 82–83
Deregulation, 3–4, 61–62, 65
  environmental issues of, 159–60
  government roles and, 127, 129–30
  See also Regulation of electricity industry; Restructuring of electricity industry
Diesel generators, 16, 18
Discrimination issues, 29,
  in vertical restructuring, 77, 78
Dispatching generators, 16–19, 66
  least-cost, 17, 114
  load balancing and, 106, 113–15
Distributed generation resources, 65, 69–70, 102, 125–26
Distribution of electricity, 2, 13, 21
  adequacy, 116, 118
  ancillary services, 19
  cross-subsidization, 76
  outages, 118–19, 121
  price structuring, 87–88
  rate regulation, 81–91
  regulation, 81–91
  reliability, 6, 120–21
  security, 116, 118, 124–25
  unbundling, 38
Divestiture
  in California, 40
  of investor-owned utilities (IOUs), 5, 24
  of generation, 78–80, 95
  in vertical restructuring, 78–80
  voluntary, 80
DSM. *See* Demand-side management programs

Eastern Interconnected System, 20, 120
Economies of scale, 64, 73, 129
Economies of scope, 73
Elasticity
  of demand, 154–55
  of prices, 96
Electric power industry functions, 13–14, 67*f*
  standards and practices for, 134–35
  See also Distribution of electricity; Generation of electricity; Marketing of electricity; Transmission of electricity
Electric power policy issues, 3–10, 187–97
Electric Reliability Council of Texas (ERCOT), 120
Electricity service providers (ESPs), 40–41
Emergency power replacement, 110, 111
  See also Reserve capacity
Emissions caps and reductions, 162, 177, 179
Energy and power definitions, 15
Energy efficiency of generation, 15
Energy-efficient appliances. *See* Demand-side management (DSM)
Energy Policy Act (EPAct; 1992), 28–29
Energy Power Act of 1935, 28
Environmental issues, 8–9, 18, 31, 134,
  in California crisis, 47–48
  environmental protection, 159–72
  green power, 31, 114, 166–67
  renewables and, 177
  See also Environmental protection and regulation
Environmental protection and regulation, 159–60
  air pollution, 160–62, 163*f*
  competitive aspects, 8–9, 178–79
  pricing issues, 162–64, 194
  regulation types, 169–70
  restructuring and, 159, 171–72
  technologies of generation and, 164–69
EPAct (Energy Policy Act), 28–29
Equity issues
  cost-benefit tradeoffs, 129–30, 191
  interstate cost differences, 9, 189–90
  low-income users support, 9, 135, 179–80
ERCOT (Electric Reliability Council of Texas), 120
ESPs. *See* Electricity service providers
Externalities, 131

Federal Energy Regulatory Commission (FERC), 27, 29, 32, 127–28, 167
  in California crisis, 47–49
  nodal pricing and, 91
  Order 888, 78, 134
  Order 2000, 78, 79, 85, 89, 125, 134, 143
  public power and, 143, 144, 146, 147–48
  reliability and, 116, 124–25
  wholesale price caps, 97–98
Federal government roles, 6–7, 8, 32, 127–32
  antitrust laws, 5, 27, 92–94, 105, 134
  congestion issues, 91

Federal government roles—*continued*
interventions, 130–32
mergers, 93–94, 105
power marketing authorities, 140–41, 145–47
public power regulation, 144
retail deregulation, 129–30
social benefits programs, 135, 176, 181, 183
statutory reforms, 132–33
stranded costs, 157–58
utilities (federal), 7, 23, 136–38, 145–48
See also Antitrust laws; Bonneville Power Administration; Federal Energy Regulatory Commission (FERC); Public Utility Holding Company Act of 1935 (PUHCA); Public Utility Regulatory Policies Act of 1978 (PURPA); Tennessee Valley Authority (TVA)
Federal Power Act of 1935, 28, 29
Federal power-marketing authorities. *See* Power marketing authorities
Federally owned utilities, 7, 23, 136–38, 145–48
Fees, 87–89
bypassability of, 40, 155–56, 181
congestion, 87, 90, 91
fixed, 87, 89
for stranded costs, 154
willingness to buy and, 88
Fixed costs, 17, 85, 87, 93
Fixed fees, 87, 89
Fossil fuels and plants, 14, 17*f*, 18
See also Coal-fired plants
Free riding, 121
Fuel cells, 69–70
Fuels and technologies for generation, 14–16, 17*f*
environmental issues, 18, 164–69
See also Coal-fired plants; Combined-cycle gas turbines (CCGTs); Hydroelectric power; Nuclear fuels and plants; Renewable power (nonhydroelectric)
Fully separated subsidiaries, 78
Functional unbundling, 38
Futures markets, 54

Gas-fired turbines. *See* Combined-cycle gas turbines (CCGTs)
Gasoline-powered generators, 16
Generation of electricity, 6, 13, 14–16
adequacy of, 6, 116, 118, 123–24
bilateral contracting, 114, 115
Generation of electricity—*continued*
capacity ownership, 24–25
deregulation, 65
dispatching and, 16–19, 113–15
distributed resources, 65, 69–70, 102, 125–26
divestiture of, 78–80, 95
efficiencies, 9, 15
entry issues, 101–2, 191–92
environmental issues, 18, 159–72
fuels and technologies, 14–16, 17*f*
generation disclosure, 183
grid's role in, 112–13
liability issues, 111
market power, 95
mergers in, 103–5
price caps, 92, 97–98
qualifying facilities, 28
regulation of, 19, 109
reserve capacity, 19, 110–13
responsibilities, 6, 106, 110–13
small-scale, 28, 125, 176, 182–83
technology mix issues, 14–16, 183
types, 14–16
See also Environmental protection and regulation
Geothermal power, 16, 18, 176
See also Renewable power (nonhydroelectric)
Gigawatt-hours (defined) 15
Global warming, 160, 161–62
Government roles, 6–7, 127–35
regional authorities, 135
standards and practices, 134–35
See also Antitrust laws; Federal government roles; State government roles
Green power (renewables-based), 31, 114, 166–67
Greenhouse gases, 159–60, 161–62
Grids. *See* Distribution of electricity; Transmission of electricity

Herfindahl-Hirschman Index (HHI), 100–1
Horizontal mergers, 98–99
benefits, 102–3
competitive effects, 101
efficiencies, 102–3
entry issues, 101–2
market definition, 99–100
relevant markets, 99–101
Hydroelectric power, 16, 50, 176
federally owned, 7, 27, 136–38
See also Bonneville Power Administration; Tennessee Valley Authority (TVA)
Incentive-based regulation, 4, 85
See also Incentive regulation
Incentive regulation, 84
competition and, 84–87
cost-of-service regulation vs., 85
rate setting by, 81, 84–87
Independent power producers (IPPs), 23
discrimination against, 29
Independent system owners (ISOs), 29, 33, 62
rate regulation and, 91
reliability and, 120
in vertical integration, 77–79
Industrial customers of electricity, 9, 22, 29–30, 188
in California power crisis, 41*f*
in Pennsylvania, 43–44
prices (U.S., 1999), 30
Interconnections, 20, 44–45, 67, 120
Internal combustion sources, 16, 17*f*
Interties, 20, 67
See also Transmission of electricity
Investor-owned utilities (IOUs), 22, 24–25, 27
public power and, 137, 141–42
reliability and, 120
stranded costs and, 150, 153
trends, 25*f*
IOUs. *See* Investor-owned utilities
IPPs. *See* Independent power producers
ISOs. *See* Independent system owners

Kilowatt-hours (defined) 15
Kyoto Protocol, 162

Least-cost dispatching, 17, 114
Liability issues, 111, 121–23
Line loss, 89
Liquidity of markets, 133
Load balancing, 5–6, 66–67, 106–8, 115
ancillary services and, 108–13
competition and, 106, 114–15
dispatching and, 106, 113–15
emergency power, 110
grid contracting, 112
load following and, 109–10
reliability and, 106–20
reserve requirements, 111–12
Load following, 19, 109–10
Load pockets, 100
Load profiles, 109
Locational marginal pricing, 90–91
Loop flows, 20, 67–68, 88
Loops (transmission), 20

Low-income users support, 9, 135, 179–80
  See also Public service programs
Lump-sum charges, 154

Marginal costs, 85, 87
Marginal pricing, 90–91
Market-clearing prices, 114
Market power, 57, 63, 77, 93, 95–96, 130, 194
  antitrust laws and, 93–94
  in California crisis, 56–59
Market share, 100
Marketers, 24, 188–89
Marketing of electricity, 13, 21, 24
  deregulation, 65
  management of, 113–15
Markets and market behavior
  advantages, 190–92
  competitive effects, 101
  concentration, 95
  liquidity, 133
  market design, 193
  market power, 56–59, 63, 93–96, 130, 194
  relevant markets, 99–101
  See also Market power
Megawatt-hours (defined) 15
Merger analysis, 101
Mergers, 5, 101–2
  competition and, 92, 93–94, 105
  competitive effects, 101, 104–5
  convergence, 98, 103–4
  efficiencies, 101–3
  horizontal, 98–103
  relevant markets, 99–101
  vertical, 103–4
  See also Antitrust laws; Horizontal mergers
Metering (real-time), 5
  in California crisis, 53–54, 57
Monopolies in electricity industry, 4, 14, 63
  natural monopolies, 63–67
  regulation of, 61–68
  vertical integration by, 74–75
  See also Mergers
Monopoly power, 63
  See also Market power
Municipal utilities, 139–40

NAERO. *See* North American Electricity Reliability Organization
National Energy Act of 1978, 28
Natural monopolies, 63–67
NERC. *See* North American Electricity Reliability Council
Net metering, 182–83
Nodal pricing, 90–91
Nonspinning reserves, 19
North American Electricity Reliability Council (NERC), 21, 116, 117–20, 124
  regions (map), 119*f*
North American Electricity Reliability Organization (NAERO), 21, 124–25
Nuclear fuels and plants, 8, 17*f*, 18
  decommissioning, 14–15, 151
  prospects, 165–66
  stranded costs of, 151, 153

Outages and blackouts, 17, 21, 117–19, 121, 195–96
  in California crisis, 46, 47*t*
  reliability and, 116–25
  See also Reliability of electric power
Ownership of generation capacity, 24–25

Pancaking of rates, 88–89
Parallel flows (loop flows), 20, 67–68, 88
PCAs (power control areas), 120
Pennsylvania and restructuring, 41–45
  California compared, 42
Performance-based regulation, 85
  See also Incentive regulation
Plant siting issues, 9
PMAs. *See* Power marketing authorities
Pollutants. *See* Air pollution
Pollution offsets, 172
Pool vs. bilateral contracting, 113–15
Postage stamp pricing, 88–89
POUs. *See* Publicly owned utilities
Power and energy definitions, 15
Power control areas (PCAs), 120
Power exchange (PoolCo), 113–14
Power loss, 89
Power marketers, 24
Power marketing authorities (PMAs; federal), 136–38, 140–41, 142, 145–48
  See also Bonneville Power Administration; Tennessee Valley Authority (TVA)
Power pools, 44, 107
Power sources. *See* Generation of electricity; Fuels and technologies for generation
Price cap regulation, 5, 84, 86
  in California crisis, 48
  in United Kingdom, 37
  wholesale, 92, 97–98
Price controls and regulation, 62–64, 68–69
  in California crisis, 48, 51–52, 57
  price caps, 5, 37, 48, 84, 86
  retail, 52, 57
  wholesale, 5, 51–52
Price takers, 85
Prices and pricing
  average (U.S., 1999), 30*f*
  in California crisis, 46–47, 97
  of distribution, 87–88
  environmental protection and, 162–64, 191
  market-clearing, 114
  nodal, 90–91
  pancaking, 88–89
  postage stamp pricing, 88–89
  price–cost margins, 96, 97
  price declines, 163–64
  provider differences, 30–31
  real-time pricing, 193–94
  restructuring and, 87–89, 150
  state differences, 30, 31*f*
  transmission, 87–89
  variations, 29–31
  willingness to buy and, 88
  See Price cap regulation; Price controls and regulation
Privatization, 35, 147–48
Profit motive, 190–91
Public-benefit programs. *See* Public service programs
Public goods, 110
Public power, 138
Public power systems, 9, 40, 136–38
  See also Cooperatively owned utilities; Publicly owned utilities (POUs)
Public service programs, 9, 40, 173–74
  cost recovery, 173, 180–81
  demand-side programs, 9, 174–76, 179
  funding, 180–81
  low-income support, 9, 135, 179–80
  renewable sources, 9, 176–77, 182
  research and development, 135, 178–79
  state roles, 178, 181, 182, 183
  support of, 180–83
Public utility commissions (PUCs), 27, 128
  renewables and, 176–77
  stranded costs and, 153
Public Utility Holding Company Act of 1935 (PUHCA), 2, 27, 28, 32, 132
  vertical integration and, 75, 80

Public Utility Regulatory Policies Act of 1978 (PURPA), 2, 8, 28, 31, 32, 65
renewable sources and, 132–33, 157, 176–77
Publicly owned utilities (POUs), 7, 22, 23*f*, 27, 136–38
in California power crisis, 40
community choice, 145
federal controls, 141–43
financial advantages, 141–43
industry functions by, 139*f*
municipal utilities, 139–40
preferential power access, 142–43
prospects, 143–45
regulatory exemptions, 143
states and, 144–45
PUCs. *See* Public utility commissions
PUHCA. *See* Public Utility Holding Company Act of 1935
PURPA. *See* Public Utility Regulatory Policies Act of 1978

Qualifying facilities (QFs), 28

Ramsey pricing, 88, 155
Rate caps, 164–65
Rate-of-return regulation, 4, 83–84, 175
Rate regulation (transmission and distribution), 81–82
incentive regulation, 84–85
price structuring, 82, 87–88
rate of return regulation, 83–84, 155
traditional, 82–83
Rate setting, 4–5, 81, 84–87
Rate structuring, 82, 87–88
Real-time pricing, 193–94
Regional reliability councils, 119–20
Regional transmission organizations (RTOs), 29, 62, 78–80, 133
functions of, 79
reliability and, 116, 120, 124–25
Regulation of electricity industry, 3–9, 26
background, 26–32
competition and, 3–5, 26–32, 61–62, 68–69
cost-of-service regulation, 82
of distribution, 81–91
incentive regulation, 4, 81, 84–87
monopoly regulation, 61–68
partial, 3–4, 61, 68–69
price issues, 6, 51–52, 57, 62–64
public power, 143, 144–45, 146
rate-of-return, 4, 81, 83–84
rate regulation (T&D), 81–91
regulatory compacts, 149, 153
stranded costs from, 149–58
Regulation of electricity industry—*continued*
of transmission, 81–91
vertical restructuring and, 75–77
See also Environmental protection and regulation; Federal Energy Regulatory Agency (FERC); Price cap regulation; Price controls and regulation; Stranded costs; Telecommunications industry regulation
Regulation of generation, 19, 109
environmental, 159–72
See also Environmental protection and regulation
Regulatory compacts, 149, 153
Reliability of electric power, 6, 9–10, 21
adequacy and, 6, 116, 118, 123–24
of bulk power, 116, 121–23
competition and, 6, 9–10, 122–23, 187, 194–97
of distribution systems, 120–21
ensuring, 118–20, 123–26, 197
load balancing and, 106–15
security and, 6, 116, 118, 125–26
systems for, 118–20
of transmission, 6, 116, 121–22
See also Load balancing; North American Electricity Reliability Council (NERC); North American Electricity Reliability Organization (NAERO)
Renewable portfolio standard (RPS), 182
Renewable power (nonhydroelectric), 9, 16, 18, 135, 176–77, 182
green power, 31, 114, 166–67
prospects, 166–67
PURPA and, 28, 132–33, 157, 176–77
stranded costs, 149, 151, 153
See also Public service programs
Research and development subsidization, 135, 178–79
See also Public service programs
Reserve capacity, 19
nonspinning reserves, 19, 110
reliability, 122
requirements, 111–12
spinning reserves, 19, 110, 122
Residential customers of electricity, 9, 22, 29–30, 188
in California power crisis, 41*f*
in Pennsylvania, 43–44
prices (U.S., 1999), 30
Restructuring of electricity industry, 1–10
in California, 38–41, 42*t*
in Chile, 34–35
Restructuring of electricity industry—*continued*
competition and, 3–9
government authority in, 128–29
issues discussed, 1–10, 24–25
market advantages, 190–92
in Pennsylvania, 41–45
policies compared, 42t
prospects, 187, 194–97
public power and, 136–38, 143–45
reliability issues, 116–25
status (U.S., 2001), 39*f*
in United Kingdom, 35–37
vertical restructuring, 71–80
See also California power crisis; Telecommunications industry regulation; Vertical restructuring
Retail electricity market, 3, 134
deregulation and government, 129–32
price controls, 51, 57, 193
sales, 22
RTOs. *See* Regional transmission organizations
Rural electric cooperatives, 22–23, 140
See also Cooperatively owned utilities

Securitization, 40, 156
Security of electric power, 6, 116, 118
coordination, 120, 124
incentives and, 125–26
See also Reliability of electric power
Sellers of electricity, 22–23
See also Cooperatively owned utilities; Federally owned utilities; Independent power producers (IPPs); Investor-owned utilities (IOUs); Publicly owned utilities (POUs)
Shopping credits, 155
Shut-down issues, 18
Small-scale generation, 28, 125, 176, 182–83
Social benefits programs. *See* Public service programs
Solar power, 16, 18, 176
See also Renewable power (nonhydroelectric)
Spinning reserves, 19, 109, 122
Spot markets, 34, 54, 124
Staged emergencies in California crisis, 46, 47*f*
Start-up issues, 18
State government roles, 6–7, 127–29, 132
in California, 46–58, 181
congestion issues, 91

State government roles—*continued*
cost difference issues, 9, 189–90
equity and tradeoffs, 129–30, 191
interstate efforts, 127, 131, 133, 135
market choices, 134
in Pennsylvania, 41–45
public power and, 144–45
public service programs, 178, 181, 182, 183
regional authorities, 135
retail deregulation and, 129–30
stranded costs, 150, 151–52, 156–57
See also California power crisis; Pennsylvania and restructuring; Public utility commissions (PUCs)
Steam turbine plants. *See* Coal-fired plants
Stranded benefits, 151, 174–76, 180
Stranded costs, 8, 40, 44, 149–51, 163
calculating, 152
in California, 39–40, 42*t*
competitive neutrality, 155–56
fees for, 154–56
issues of, 153–54
levels of, 151–52
in Pennsylvania, 42*t*, 44
politics and, 157–58
Subsidies
consumers, 174–76, 179–80
cross-subsidization, 75–76, 80, 95, 121
demand-side, 174–76
renewables, 176–78
research and development, 135, 178–79
Supply issues. *See* Distribution of electricity; Generation of electricity

T&D (transmission and distribution), 82
regulation of, 81–91
See also Distribution of electricity; Transmission of electricity
Telecommunications industry regulation, 3, 4, 61, 66, 67, 188–89
AT&T breakup, 72, 75–76, 80
incentive regulations, 84, 86–87
Telecommunications Act of 1996, 71, 80
Tennessee Valley Authority (TVA), 136, 138, 140–41, 146–47
Texas Interconnected System, 20
Transco (transmission company), 78–79
Transmission loading relief procedures (TLRs), 122–23
Transmission of electricity, 2, 13, 19–21
adequacy, 116, 118
bulk power, 116, 121–22
congestion, 20, 87, 89–90, 91
control areas, 20–21
costs, 19–20
cross-subsidization, 76
dispatching and, 16–19, 113–15
distributed generation and, 125–26
failure, 19
generation and, 112–13
grids (interties), 19–21, 133
interconnections, 20, 44–45, 67, 120
loops, 20
monopoly, 67–68
power pools, 20–21
pricing issues, 27, 87–89
rate regulation, 81–91
regulation, 81–91
reliability, 6, 116, 121–22
RTOs, 29, 62, 78–80, 116, 120, 124–25, 133
security, 116, 118, 124–25
technology issues, 19–20
transmission loading relief procedures (TLRs), 122–23
transcos, 78–79
unbundling, 38
See also Regional transmission organizations (RTOs)
TVA. *See* Tennessee Valley Authority
Two-part tariffs, 88

Unbundling
functional, 38
of prices, 40
United Kingdom and restructuring, 35–37
Utilities, 9, 22–23
See also Cooperatively owned utilities; Federally owned utilities; Independent power producers (IPPs); Investor-owned utilities (IOUs); Publicly owned utilities (POUs)

Vertical "convergence" mergers, 98, 103–4
Vertical integration, 13–14, 72–74
ISOs in, 77–79
RTOs in, 78–79
See also Vertical restructuring
Vertical restructuring, 71–80
cross-subsidization, 75–76, 80
divestitures in, 78–80
economies of scope, 73
monopolies in, 74–75
regulation, 75–77
vertical integration, 72–74
Voltage control, 19

Watt-hours (defined) 15
Western Interconnected System, 20, 120
Wheeling, 27, 69
Wholesale electricity market, 22, 134
price caps, 92, 97–98, 194
Willingness to buy/pay, 88
Wind power, 16, 18, 176
See also Renewable power (nonhydroelectric)